STRANGE SCIENCE

BY THE EDITORS OF PORTABLE PRESS

PORTABLE PRESS
SAN DIEGO, CALIFORNIA

Portable Press
An imprint of Printers Row Publishing Group
10350 Barnes Canyon Road, Suite 100, San Diego, CA 92121
www.portablepress.com
e-mail: mail@portablepress.com

Printers Row Publishing Group is a division of
Readerlink Distribution Services, LLC.
Portable Press is a registered trademark of
Readerlink Distribution Services, LLC.

All correspondence concerning the content of this book should be
addressed to Portable Press, Editorial Department, at the above
address.

Publisher: Peter Norton
Publishing/Editorial Team: Vicki Jaeger, Tanya Fijalkowski,
Lauren Taniguchi, Aaron Guzman
Editorial Team: JoAnn Padgett, Melinda Allman,
J. Carroll, Dan Mansfield
Production Team: Jonathan Lopes, Rusty von Dyl

Library of Congress Cataloging-in-Publication Data

Names: Portable Press (San Diego, Calif.)
Title: Strange science.
Description: San Diego, CA : Portable Press, [2017]
Identifiers: LCCN 2017001634 | ISBN 9781626869820 (pbk.)
Subjects: LCSH: Science--Miscellanea. | Science--Popular works.
Classification: LCC Q173 .S8887 2017 | DDC 500--dc23
LC record available at https://lccn.loc.gov/2017001634

Printed in the United States of America
First Printing
21 20 19 18 17   1 2 3 4 5

STRANGE SCIENTISTS

J. Carroll

Lidija Tomas

Michael Sherman

Sophie Hogarth

Dan Mansfield

Vicki Jaeger

JoAnn Padgett

Jay Newman

Tanya Fijalkowski

Rusty von Dyl

Anna Nguyen

Introduction

From archaeology to zoology, from alien DNA to X-ray guns, this epic edition from the editors at Portable Press delves into the weirdest of weird science. While combing through the strangest discoveries from days of yore through the modern age, we found that science can be used to do almost anything: solve history's mysteries, cure diseases, get better at sports, increase the world's food supply, and make people's lives easier...or shorter (cue ominous music). *Strange Science* has all this and more—it will answer questions about the human body, correct common misconceptions, arm you with scientific lingo, reveal the truth behind bad science, and even teach you a little about quantum physics!

BONUS: If you're not content to be an armchair scientist, you'll also engage in quizzes and discover "realistic" do-it-yourself projects, including how to hypnotize a chicken or make a mummy. So put on your thinking cap, or your tinfoil hat to block out mind-control waves—depending on which branch of science you prefer—and get ready for an experimental journey!

THE CURE FOR WHAT AILS YE

We begin by traveling back to a time before modern medicine, when these "scientific" remedies were recommended.

To prevent consumption (tuberculosis):
"Let not your breast touch the table or desk on which you write, for leaning the breast hard against the edge of the table hath brought many young men into a consumption."
—*The Young Man's Companion* **(1775)**

For alcoholism:
"The prescription is simply an orange every morning a half hour before breakfast. Take that and you will neither want liquor nor medicine. The liquor will become repulsive."
—*Dr. Chase's Recipes* **(1884)**

To prevent influenza:
"Children should be instructed to run with the mouth shut for the first block or two after going outdoors in cold weather."
—*The Guide Board to Health, Peace and Competence* **(1870)**

To treat asthma:
"A pipe of tobacco (or a cigar) has the power of relieving a fit of asthma, especially in those not accustomed to it."
—*Cassell's Household Guide* **(1880)**

To treat epilepsy:
"It has been said that a black silk handkerchief, thrown over the face while the fit is on, will bring the person 'to' instantly."

—*The Guide Board to Health,*
Peace and Competence (**1870**)

To cure stuttering:
"Let him who stammers, stamp his foot on the ground at the same time that he utters each syllable and stammering is impossible."

—*Fun Better than Physic* (**1877**)

"Nothing is better than ear-wax
to prevent the painful effects resulting from a wound by a nail, skewer, etc. It should be put on as soon as possible. Those who are troubled with cracked lips have found this remedy successful when others have failed."

—*The American Frugal Housewife* (**1832**)

ROBOTS: NOW WITH A TASTE FOR FLESH

Scientists at NEC System Technologies in Japan have invented a robot that can taste and identify dozens of wines, as well as some types of food. The green-and-white tabletop robot has a swiveling head, eyes, and a mouth that speaks in a child's voice. To identify a wine, the unopened bottle is placed in front of the robot's left arm. An infrared beam scans the wine—through the glass bottle—and determines its chemical composition. The robot then names the variety of wine, describes its taste, and recommends foods to pair it with. Scientists are still working out the kinks: At a press conference, a reporter and a cameraman put their hands in front of the robot's infrared beam. According to the robot, the reporter tasted like ham, and the cameraman tasted like bacon.

THE SCIENCE BEHIND KIDS' PRODUCTS

SILLY STRING

It's not really the "string" that makes Silly String work; it's the ingredient that gets it out of the can that makes it all happen. The strands are created from an acrylic resin—plastic—and a surface-acting agent—foam. But it's the propellant that's crucial to the process: Not only does it push the string out of the can when sprayed, it also causes the reaction between the resin and the surface-acting agent to form the sticky strands.

ELMER'S GLUE–ALL

Although many glues were traditionally manufactured from the collagen in animal hooves, horns, and bones, Elmer's never has been. The first glue factory that Gail Borden bought in 1929 used a milk by-product to make its glue, and Borden Inc. expanded into resin-based glues in the 1930s. Elmer's Glue-All was introduced in 1947 and was made from a synthetic resin, which is still in their product today. So kids can play happily in the knowledge that no horses were harmed in the making of their Elmer's.

RISE OF THE FUTURISTS

For most of human history, if you sought advice from a shaman, a soothsayer, or Nostradamus, you'd hear whatever the bones or the crystal ball "told" them. In the mid-1400s, the advent of the printing press (and with it the book industry) made the world's accumulated knowledge available to the masses (at least to the ones who could read). That advance ushered in the Age of Enlightenment, followed by the Industrial Revolution. People began to look at the future from a more scientific point of view.

The first futurists weren't necessarily scientists, but had a keen understanding of both history and human nature. That concept is called *foresight*. "It refers to a process of visioning alternative futures through a combination of hindsight, insight, and forecasting," explains Tuomo Kuosa in his book *The Evolution of Strategic Foresight*. "(Hind)sight is about systematically understanding the past, (in)sight is about systematically understanding the true nature of the present, and (fore)sight is about systematically understanding the future."

One of the first men to display that foresight was Irish satirist Jonathan Swift. In his 1726 novel *Gulliver's Travels*, the hero travels to a strange island full of futuristic gadgets—one of them a giant "Engine" containing "Bits" that allow even "the most ignorant Person to write Books in Philosophy, Poetry, Politicks, Law, Mathematicks, and Theology." It's all "linked

9

together by slender Wires." Swift basically described electricity, computers, and the Internet hundreds of years before they were invented.

Even more impressive, Swift wrote about "two lesser stars, or satellites, which revolve around Mars." How did he know that Mars had two moons 150 years before they were discovered? He wasn't psychic (as some assumed), just logical: the two planets closest to the Sun have no moons, ours has one, and it was known even then that the large outer planets have several moons. Mars, Swift concluded, would most likely have two. His foresight was spot-on.

10 CLOSEST STARS TO EARTH

1. The Sun
   (0.000016 light-years)
2. Proxima Centauri
   (4.2 light-years)
3. Alpha Centauri A
   (4.3 light-years)
4. Alpha Centauri B
   (4.3 light-years)
5. Barnard's Star
   (5.96 light-years)
6. Wolf 359
   (7.6 light-years)
7. Lalande
   (8.11 light-years)
8. Alpha Sirius
   (8.7 light-years)
9. Beta Sirius
   (8.7 light-years)
10. A Luyten
    (8.93 light-years)

TRUE TV SCIENCE

Game of Thrones (2011–). This fantasy series features surreal weapons and unbelievable ways of killing off our favorite characters, but surprisingly, some of it is plausible. Season Six mentions a mystical sword made of meteorite. Ancient Egyptians, before they had the technology to smelt iron, crafted swords and daggers from meteoritic iron. King Tutankhamun, Mughal Emperor Jahangir, and Attila the Hun had such weapons. And when the character Khal Drogo kills his brother-in-law Viserys Targaryen by crowning his head in molten gold, it recalled natives of South America who poured hot gold down the throat of a gold-taxing Spanish governor in 1599. (The fire over which Drogo heats the gold would have to be an impossible 1,947 °F, but if he added lead to the mixture, that would lower its melting point and make the storyline feasible.)

The Walking Dead (2010–). In this postapocalyptic series, dead people are reanimated in the form of uncoordinated, confused zombies. In reality, Haitian voodoo priests have apparently turned people into the "walking dead." They administer a nerve toxin from pufferfish that brings the victim to the brink of death, paralyzed and smelling rotten. Then the witch doctor may

use a poisonous plant called angel's trumpet to "resurrect" them, leaving them delirious and sometimes brain-damaged, yet able to walk.

Penny Dreadful (2014–2016). Considering the show's werewolves, witches, and other fictional characters, it's hard to believe that some of its storylines may have a factual basis. The character of Dr. Victor Frankenstein, for instance, uses dead bodies he finds via the black market to conduct experiments with electricity, ultimately bringing them back to life. In the 1790s, Italian doctor Luigi Galvani electrified dead animals to make them twitch—and in doing so, pioneered the creepy-sounding field of electrophysiology. And resurrectionists did steal freshly dead bodies from graveyards to deliver them to medical "professionals."

NOT AS GREAT AS IT SOUNDS

Sexsomnia is an actual medical condition in which a person performs a sexual act while asleep. Sleep sex has even resulted in criminal charges being brought against sleeping people who had sex with someone who wasn't willing, or with a minor, while being completely unaware of it.

ANTARCTIC JARGON

Living as a research scientist at the McMurdo Station in Antarctica must be a unique experience. And it has its own lingo.

BOOMERANG: An outgoing airplane flight that has to return immediately after takeoff due to bad weather.

GREEN BRAIN: A small green notebook issued to all researchers.

IVAN: Short for "ice van," it's the large, iceworthy bus that transfers researchers from one building to another.

APPLES: Warming huts—red, fiberglass, domed igloos.

POLIE: Research workers (as in "South Polies").

ODEN: Named after the powerful Norse god Odin, it's a huge icebreaking vessel used on the water channels where the supply ships enter.

GERBIL GYM: The workout room, which consists almost entirely of treadmills.

FRESHIES: The weekly food delivery from New Zealand.

SOUTHERN: There are two bars for the scientists off duty. The Southern (short for Southern Exposure) allows smoking; the other bar, Gallagher's, doesn't.

WINFLY: The day-long switchover from the winter crew (Feb.–Oct.) to the summer crew (Oct.–Feb.). It's short for "winter fly-in."

YAK TRACKS: Traction-providing grips that go on the bottom of boots.

BIG RED AND BUNNY BOOTS: The two main pieces of standard issue ECW, or "extreme cold weather" gear. Big Red is a big puffy coat; Bunny Boots are white rubber boots.

UPPERCASE: The three-story dormitory researchers live in.

THE ICE: Antarctica itself.

FNG: Pronounced "fingee," it means a new person on the station. (NG stands for "new guy." You'll have to guess what the F stands for.)

BIOHACK U

Biohackers don't hack computers; they hack their own bodies, looking for ways to upgrade their bodies using technology. "We hack our bodies with artifacts from the future-present," states one biohacking website. Good idea? Well, Dr. Anthony Guiseppi-Elie, professor of bioengineering at Texas A&M University, says, "Anyone doing this should stop!"

IMPLANTABLE COMPASS

Southpaw, designed by electronic engineer and biohacker Brian McEvoy, is a miniature compass to be implanted under your skin. An ultrathin whisker sticks out of Southpaw's rounded titanium shell. When you face north, the whisker tickles the underside of your skin. "It would be best located near the shoulder," says McEvoy, who plans to be the guinea pig for testing his own device.

INTERNAL EARPHONES

Rich Lee, 34-year-old salesman and biohacker, wears sound-transmitting magnets implanted in his outer ears and a wire coil around his neck that converts sounds into electromagnetic fields. Those fields thus become "internal headphones." With a media player, amplifier, and battery pack hidden under his shirt, Lee can listen to music all day long with no one the wiser. He can also "hear" heat from a distance and detect magnetic fields and Wi-Fi signals.

How to Hypnotize a _Chicken_

If you ever get a chance to place a chicken under your spell, give it a try—it's fascinating to watch, and harmless and painless for the chicken. (Who knows— you might even win a bar bet.)

STEP 1. Techniques vary widely from place to place. Some methods call for laying the chicken gently on its side, with one wing under its body, holding it in place with one hand so that your other hand is free. Others say that turning the chicken upside down, lying on its back with its feet up in the air, is best. Either way, the disoriented bird will need a second to regain its bearings, but once it does it will not be bothered by being in this unfamiliar position.

STEP 2. Some hypnotists advocate placing a finger on the ground at the tip of the chicken's beak and drawing a line four inches long in the dirt extending out from the beak and parallel to it (picture Pinocchio's nose growing). Trace your finger back and forth along the line for several seconds. Other practitioners say that drawing a

circle, not lines, in the dirt around the chicken's head works best. Still others say all you need to do is stroke the chicken on its head and neck with your index finger. If one method doesn't seem to work, try another.

STEP 3. Whichever method you try, keep at it for several seconds. That's about how long it takes for a chicken to go into a trance. Its breathing and heart rate will slow considerably, and its body temperature may even drop a few degrees.

STEP 4. You can now let go of the chicken. It will lie perfectly still in a trancelike state for several seconds, several minutes, or even an hour or more before it comes out of the trance on its own. You can also awaken the chicken yourself by clapping your hands or nudging it gently. (The unofficial world record for a chicken trance: 3 hours, 47 minutes.)

STEP 5. If holding a chicken in one hand while hypnotizing it with the other proves too difficult, another technique calls for putting the chicken in the same position it goes into when it's asleep—with its head under one wing—and rocking it gently to induce a trance.

CHICKEN SCIENCE

Just as there are different theories as to which method of chicken hypnotism is best (see previous page), so too are opinions divided as to what exactly is going on with the chicken when it is being hypnotized.

- The trance could be a panic "freeze" response, similar to a deer stopping in the middle of the road when it sees headlights.

- It may also be an example of *tonic immobility*, a reflex similar to an opossum's ability to go into a trancelike state when it feels threatened. Chickens roost in the branches of trees or other high places at night; the trance reflex, if that is indeed what it is, may help the chicken to remain perfectly still, silent, and (hopefully) unnoticed as foxes, raccoons, and other predators prowl below.

Used-Less Invention

GRAVITY-POWERED SHOE AIR CONDITIONER

PATENT NUMBER: 5,375,430

INVENTED IN: 1994

DESCRIPTION: The air-conditioned shoe can either cool your foot or warm it up, depending on your preference. Hidden inside the shoe's heel are expanding and compressing chambers powered by the natural pressures that occur while walking. With each step, networks of heat-exchange coils work with the chambers to alter the temperature of the air surrounding your foot. End result: A sweat-free (but bulky and cumbersome) shoe.

"The art challenges the technology, and the technology inspires the art."

—John Lasseter

"Science" Museums

These places define "science" and "museum" loosely.

BARBED WIRE MUSEUM

Location: La Crosse, Kansas

Background: Sure, barbed wire is an important part of American history. It provided an inexpensive way for "sodbusters" to keep cattle off their land, effectively ending the open range. But wire's wire, right? Apparently not. This museum holds 18-inch segments of more than 1,000 different types of barbed wire, lining its walls from floor to ceiling. Don't miss the real bird's nest made almost completely from bits of barbed wire (it weighs 72 pounds) and a piece of barbed wire from the top of the Berlin Wall.

SPAM MUSEUM

Location: Austin, Minnesota

Background: Next door to the top-secret facility where the Hormel Corporation makes Spam, fans of the canned meat can see a giant Spamburger sandwich (with its own 17-foot spatula), visit the 3,390-can "Wall of Spam," don hardhats and work on a simulated Spam production line, and marvel at the 4,700 cans that document 70 years of Spam's worldwide popularity. Check out the talking wax figure of company founder George Hormel.

EXPENDABLE ORGANS

APPENDIX

Only some mammals have an appendix—rodents, rabbits, marsupials and primates (including humans). Located near the junction of the small and large intestines, it is a no-longer-useful stump of a much larger pouch.

WHAT IT'S FOR: The appendix was once part of a larger cellulose-digesting pouch left over from ancient times when humans were mostly herbivores.

WHY WE CAN DO WITHOUT IT: Although some scientists have recently speculated that the appendix might carry a reservoir of useful gut microorganisms, humans can certainly live without it. When an appendix gets infected and bursts, the spread of toxic fluid can kill the patient. It's also possible to get appendix cancer.

SIDE EFFECTS OF REMOVAL? A small number of patients may develop infections or have reactions to anesthesia, but compared to the risks of not taking out an infected appendix, the risks are pretty low.

Mystery Manuscript

When Wilfrid Voynich purchased a strange, antiquated manuscript from Roman Jesuits in Italy in 1912, he was fascinated by its mysterious drawings and elegant script in a language that no one had ever seen before. He took his so-called cipher manuscript to countless cryptography experts—all of whom have failed to crack the code even to this day. The nearly 250 pages are filled with illustrations of unknown plants, naked women swimming in what appear to be glass tubes, and zodiac signs in odd arrangements, all accompanied by handwritten text no one can decipher.

For hundreds of years, no one has been able to explain the manuscript's origins and meaning. A cover letter found inside the manuscript speculated that the author was Roger Bacon, a 13th-century English friar and natural philosopher. Other candidates include John Dee, a 17th-century mathematician and philosopher, or his associate Edward Kelley, who may have made the manuscript as a hoax in order to defraud the Bohemian emperor Rudolf II. However, radiocarbon dating of the paper shows that it was probably produced in the early 15th century, which eliminates all three men from the running.

Others suggest that the manuscript was the work of a medieval charlatan who was trying to impress a client, or that it is simply gibberish written by a crazy person.

Still others claim the writer was channeling teachings from spirits or even aliens from the Pleiades stars. The true author of the Voynich manuscript may never be unveiled, but a facsimile of the book is available online if you think you've got what it takes to solve a puzzle that has flummoxed the world's finest cryptographers for centuries. (Sure, you do...)

Scientists have figured out how to turn sugar—cheap, plentiful sugar—into a lab-created substance called graphene. It's now the single strongest substance in the world, yet it's also completely flexible. Produced in sheets of treated carbon that are just one atom thick, graphene may be used in touchscreens, lighting, and much more.

DUMB PREDICTIONS

"I predict the Internet will
soon go spectacularly
supernova and in 1996
catastrophically collapse."

—ROBERT METCALFE,
INVENTOR OF ETHERNET, 1995

STRANGE MEDICAL CONDITION

SUBJECT: A 49-year-old man in Brazil

CONDITION: "Pathological generosity"

STORY: In August 2013, the medical journal *Neurocase* published a report by Brazilian doctors about a 49-year-old man, referred to only as "Mr. A," who had undergone a bizarre personality change after suffering a stroke. The change: he could not stop himself from giving things to people. That included buying candy, food, and gifts for kids on the street and giving away his money to people he hardly knew. The man was so prone to "pathological generosity," the report said, that over the course of just a few years his behavior put a serious financial strain on his family, especially after it caused him to lose his job as a manager at a large corporation. The doctors said they believed the bizarre symptoms were caused by damage the stroke did to an area of the man's brain (the subcortical region of the frontal lobe—for you neuroscientists playing along at home) that is known to take part in the regulation of human behavior. They added that Mr. A's was the only known case of "pathological generosity" ever recorded.

The Hand of Glory

If you're familiar with Harry Potter, then you remember the Hand of Glory, a magic weapon that Draco Malfoy acquires. Strangely enough, back in the Middle Ages, the Hand of Glory was considered actual science by some. Here's how it (supposedly) worked: When an executed criminal was still hanging from the gallows, a burglar would sneak up and cut off the dead man's right hand. Later the burglar would drain the blood from the hand and wrap it in a piece of cloth. The hand was "pickled in salt, and the urine of man, woman, dog, horse and mare; smoked with herbs and hay for a month; hung on an oak tree for three nights running, then laid at a crossroads, then hung on a church door for one night." The final step: dip the pickled appendage in fat (best to use the corpse of a criminal) and voilà—the Hand of Glory is ready.

When the burglar broke into a home, he was supposed to light the hand like a candle and recite a verse, after which the home's dwellers would fall into a deep sleep so he could burgle the night away in peace. Here it is:

> *Let those who rest more deeply sleep,*
> *Let those awake their vigils keep,*
> *O Hand of Glory, shed thy light,*
> *Direct us to our spoil tonight.*

NYE'S BALLET SHOES

Here's a strange patent from our
"Dressed By Geniuses" files.

SCIENTIST: Bill Nye

PATENT NO. US 6895694 B2: "Toe Shoes"

STORY: The "Science Guy" holds the patent for a toe shoe with a special "toe box" that's "capable of providing support to a ballet dancer's foot, toes, and ankle during en pointe dancing." Why did Bill Nye invent a better ballet shoe? Because he's a scientist who saw a problem (while filming a segment for his TV show at a Seattle ballet theater) that he knew he could fix: "These women, they're 22 years old, and they have three or four surgeries already...The toe shoe has not changed in centuries. So I just got to thinking about it." (Nye was awarded the patent, but it's uncertain if any of his toe shoes have been manufactured.)

Famous Fetus

For nine years, John and Lesley Brown of Bristol, England, had been trying to have a child, but Lesley's Fallopian tubes were blocked, thus preventing a pregnancy. In November 1977, Lesley underwent in vitro fertilization (IVF), in which an egg was extracted from her uterus, placed in a lab dish, and fertilized with John's sperm. Two and a half days later, after cells of the egg divided to become an eight-celled embryo, the egg was carefully replaced in Lesley's uterus.

IVF had been tried before, but had never resulted in a pregnancy that lasted more than a few weeks. When Lesley's pregnancy lasted months, the media went crazy over the possible birth of the first "test-tube baby." The relentless British press forced the Browns into hiding.

On July 25, 1978, at 11:47 p.m., Louise Joy was born via Cesarean section; she weighed 5 pounds 12 ounces. As the world's first baby conceived outside of the womb, Louise was a scientific milestone. Religious and ethical controversies over IVF continued to rage, but that didn't stop infertile couples from turning to the new procedure— especially since the blond, blue-eyed Louise was a normal, healthy baby.

Realistic Robots

It seems only a matter of time before robots are indistinguishable from humans—these machines are resembling more and more the T-800 from the Terminator *series.*

STRANGE HOTEL

The Henn-na Hotel (translation: "Strange Hotel") made world headlines when it opened in 2015 in Nagasaki, Japan. Why? Its staff consists of 80 robots that greet hotel guests, carry their bags, and serve food in the café. While the humanoid robot receptionists welcome you with a smile to make you feel at ease, they speak only Japanese. If you want to speak English, you'll have to approach the less-welcoming Velociraptor robot wearing a bow tie. There are about 10 human employees behind the scenes to make sure everything runs smoothly—so humans aren't replaceable...yet.

YOU'VE GOT A FRIEND IN ME

Pepper is a social robot that is programmed to recognize different facial expressions, gestures, and voices in order to make conversation with humans. All these features make Pepper an ideal companion for the elderly, home-bound patients, or lonely types who live in their mother's basement.

Strange Movie Science

And by "strange" we mean "wrong."

The Core (2003) The planet's core suddenly stops spinning! Oh no! One problem (of several): If the core stopped moving, it wouldn't take several weeks for the effects to be felt. Quite the opposite—an incredible amount of energy would be released, resulting in a massive worldwide earthquake that would lay waste to everything in a few minutes and wouldn't stop shaking for years.

Jurassic World (2015) The filmmakers of the fourth installment in the series ignored an important scientific discovery—made after 1993's *Jurassic Park*—that many smaller dinosaurs had feathers. So instead of depicting Velociraptors as turkey-sized, feathered beasts, they're still as tall as men and covered in scales.

Interstellar (2014) Cooper (Matthew McConaughey) visits a planet that's orbiting a black hole, but the planet has heat and light...which usually requires a star.

Star Trek (2009) The old version of Spock (Leonard Nimoy) speaks of a supernova—an exploding star—that would "threaten the galaxy." Sorry, Spock, but even though supernovas are big on a planetary scale, on a galactic scale—we're talking hundreds of millions of stars—one star exploding wouldn't do much damage to anything more than about 50 light-years away.

UNDER THE INFLUENCE

What goes on in your body when you've been drinking alcohol? Here are some basic facts:

1. When you drink an alcoholic beverage, your body absorbs about 90% of the alcohol in the drink. The rest is exhaled, sweated out, or passed out in urine.

2. On average, a normal liver can process 10 grams of alcohol per hour. That's the equivalent of one glass of wine, half a pint of beer, or one shot of 80 proof spirits. (Exactly how much depends on a number of things, including your weight and gender.)

3. Alcohol is a depressant, which means that it slows down the activity of your central nervous system by replacing the water around the nerve cells in your body.

4. Alcohol also changes the density of the fluid and tissue in the part of your ears that controls your sense of balance. That's why it can be difficult to walk, or even stand up, when you've had too much to drink.

STRANGE STUDY:
The Tetris Effect

STUDY: Do you have uncontrollable cravings for sex, drugs, and alcohol? Now you can control them with Tetris! So says a 2015 study conducted by researchers from England and Australia. They instructed 31 college students to log their daily cravings. The researchers sent some of the students text messages prompting them to play the block-stacking video game Tetris for three minutes, seven times a day.

CONCLUSION: "Playing Tetris decreased craving strength for drugs, food, and activities from 70% to 56%," reported Jackie Andrade from Plymouth University in the United Kingdom. "We think the Tetris effect happens because craving involves imagining the experience of consuming a particular substance or indulging in a particular activity. Playing a visually interesting game like Tetris occupies the mental processes that support that imagery; it is hard to imagine something vividly and play Tetris at the same time." So that's good news for drug and sex addicts...but not so good news for video game addicts.

WHEN A GRIZZLY LOVES A POLAR

A "grolar" is the offspring of a grizzly bear and a polar bear. Until recently, wild grolars were considered a legend. The chances of two such bears meeting, let alone breeding, are very unlikely. Brown bears prefer temperate forests whereas polar bears live in cold climates. Their "bedroom habits" are also different. Brown bears mate on land; polar bears prefer ice floes.

In 2006 an American hunter named Jim Martell teamed up with Roger Kuptana, an Inuit tracker, for an expedition on Banks Island in the Canadian Arctic. Martell shot and killed what he thought was a polar bear. Upon further inspection, he and Kuptana realized what they had on their hands was infinitely more rare. It had brown patches in its white hair and a humped back like a grizzly.

DNA evidence confirmed that their bear was, indeed, the first grolar ever discovered in the wild. Scientists still consider them to be exceptionally rare in nature, but the likelihood of a population boom could increase as grizzlies are driven farther north by civilization and as polar bears are driven farther south by global warming.

SKATEBOARD SCIENCE

In the 1960s, bored surfers started attaching roller-skate wheels to wooden boards so they could sidewalk surf when there were no waves. Several ideas changed skateboarding from a way to get around on pavement into a way to defy gravity and fly through the air.

- One early improvement to the skateboard was the kicktail, the board's upturned back end. It added a way to brake, a higher level of control, and allowed the skateboarder to lean back more as he rolled along. In the 1970s, shorter boards made of lighter materials and urethane wheels provided a smoother and quieter ride. Soon kids who'd never seen the ocean were zipping down hills and maneuvering around obstacles on skateboards.

- In 1977 a skateboarder named Willi Winkel was riding down a standard quarter pipe (an elevated ramp that led downhill to help a rider pick up speed). Winkel thought that two quarter pipes might be better than one, so he put together a U-shaped ramp or "half-pipe." He was using the rules of acceleration and velocity to overcome gravity. His total mass (weight) was pulled by

gravity down the half-pipe, thus creating speed and giving him the momentum to take him vertically up the other side and even soar out over the lip to "catch some big air" (in surfer-speak).

- In the late 1970s, Alan Gelfand worked on a move he'd learned from friend Jeff Duerr. As Gelfand sped up the vertical incline of a half-pipe, he made a crouching jump while shoving down the kicktail of his board with his back foot, deliberately torqueing the back of his board down and causing the front of the board to fly up as the back bounced off the ground. First called "due air" after its originator Duerr, it became popularized by Gelfand and later known by his nickname "Ollie." Remember the seesaw? He had taken advantage of the effects of rotational motion. By itself, the board would simply have flipped over backward toward its axis, but eventually, while the board was in the air, Gelfand learned to slide his front foot forward, which put torque on the front of the board and leveled it out before gravity pulled rider and board back to earth. Spectators were amazed; it looked as if Alan's skateboard was strapped to his feet—but it wasn't.

SCIENTIFIC STREETS

Many of the streets in Paris are named for famous scientists. Here are five you might recognize:

1. RUE AMPÈRE. Named for French physicist André-Marie Ampère, who discovered electromagnetism. He initiated a standard system of measurement for electric currents, and the ampere unit of electric current was named for him.

2. RUE COPERNIC. Named for Polish astronomer Nicolaus Copernicus, who produced a workable model of the solar system with the Sun in the center in the 16th century.

3. RUE PIERRE ET MARIE CURIE. Named for the Nobel Prize–winning couple who pioneered the study of magnetism and radioactivity, and discovered the elements radium and polonium in 1898. (Polonium was named for Marie's homeland of Poland.)

4. RUE GALILÉE. Named for Italian physicist, mathematician, astronomer, and philosopher Galileo Galilei, who has been called the "father of modern science."

5. RUE FOUCAULT. Named for Jean Bernard Léon Foucault, a French mathematician and astronomer who invented the gyroscope and a pendulum that demonstrated that Earth rotates on its axis.

THE MUMMIES RISE

As long as there have been people in Egypt, there have been mummies—not necessarily man-made mummies, but mummies nonetheless. The extreme conditions of the desert environment guaranteed that any corpse exposed to the elements for more than a day or two dried out completely, a process that halted decomposition in its tracks.

The ancient Egyptian culture that arose on the banks of the Nile River believed very strongly in preserving human bodies, which they believed were as necessary a part of the afterlife as they were a part of daily life. The formula was simple: no body, no afterlife—you couldn't have one without the other. The only problem: As Egyptian civilization advanced and burial tombs became increasingly elaborate, bodies also became more insulated from the very elements—high temperatures and dry air—that made natural preservation possible in the first place.

So a new science emerged: artificial mummification. From 3100 B.C. to A.D. 649, the ancient Egyptians deliberately mummified their dead, using methods that became more sophisticated and successful over time.

For more about mummies, go to page 208.

MANIMALS!

Stanford University professor Irving Weissman and a team of researchers have created mice with brains that are part human. Hoping to learn more about brain cancers, Weissman extracted human embryonic brain stem cells—the kind that go on to become various types of brain cells—and injected them into the brains of adult mice. The cells survived and even traveled to different areas of the brains and matured into different types of brain cells. (The researchers created special markers that allowed them to keep track of the injected human cells.) The tests resulted in mice with brains whose cells were about 1 percent human. The next step: inject human brain stem cells not into *adult* mice but into fetal mice still in the womb. That, Weissman says, would result in mice that have much higher human brain content...perhaps as much as 100 percent. Before moving ahead, Weissman went to Stanford's ethics department to make sure he wasn't crossing any lines. Law professor Henry Greely, chair of the school's ethics committee, gave the study the go-ahead with one condition: If the mice started showing any humanlike behaviors, they'd have to be destroyed immediately.

AN UNFAMILIAR FACE

There's a part of your brain that processes faces. It's located, according to MIT scientist Nancy Kanwisher, in the area "just behind and underneath, and a bit from your right ear." It's called the fusiform gyrus. (The gyrus is a ridge in the brain, and fusiform describes its shape—elongated and tapered at both ends.) Whenever you see someone you know, the fusiform gyrus tells you, "That's Bob." It also sends out messages to other parts of the body that add emotions to the information, such as "I like Bob. He's my friend." But what happens when an accident, illness, or hereditary gene disconnects the wiring between the fusiform gyrus and other parts of the brain?

There are people who may see a particular person's face every day of their lives and still not recognize it. They see a nose, teeth, and cheeks, but when the features are put together, they cannot retain a memory of it. The medical term for this condition is *prosopagnosia* (from the Greek *prosopon,* for "face," and *agnosia,* for "ignorance"), but it's more commonly called *face blindness*. Researchers say that as many as 1 in 50 people suffer from some form of the condition.

Jane Goodall, the world's foremost expert on chimpanzees, has it. Probably the best known sufferer of prosopagnosia is the neurologist and psychiatrist Dr. Oliver Sacks, renowned author of the best-selling books *The Man Who Mistook His Wife for a Hat* and *Awakenings*, which was made into the 1990 Oscar-nominated film starring Robin Williams. A lifelong sufferer of extreme face blindness, Sacks has said that his condition is so severe he often doesn't recognize his own face.

Sufferers of face blindness must develop alternate ways of identifying coworkers, friends, and family, so they remember single features—a mole, a specific style of clothing, or an extra toothy smile. Says Jane Goodall, "I usually make up for it by pretending to recognize everybody. And then, if they say, 'But we haven't met before,' I say, 'Well, you look just like somebody I know.' "

---

4 DIRTY-SOUNDING SCIENCE WORDS

1. **Mastication** (chewing)
2. **Coccyx** (tailbone)
3. **Hyperprosexia** (a preoccupation with, or ability to concentrate on, only one thing)
4. **Vomitus** (yes, this is what it sounds like—the technical term for the stuff you vomit up)

YOU MUST REMEMBER

These memory tricks (or mnemonic devices*)
will help you remember scientific facts.*

THREE SEGMENTS OF AN INSECT'S BODY:
Picture a bug wearing a **hat**—**h**ead, **a**bdomen, **t**horax.

BIOLOGICAL GROUPINGS:
Kind **p**igs **c**are **o**nly **f**or **g**ood **s**lop.
(kingdom, phylum, class, order, family, genus, species)

TYPES OF CAMELS:
Bactrian (two humps), with a back that looks
like a *B* turned on its side. The one-humped
dromedary's back looks like a *D*.

**8 PLANETS IN EARTH'S SOLAR SYSTEM
(from the closest to the farthest from the Sun):**
Many **v**ery **e**vil **M**artians **j**ust **s**howed **u**p **n**aked.
(Mercury, Venus, Earth, Mars, Jupiter,
Saturn, Uranus, Neptune)

WORLD'S LONGEST RIVERS:
Just say **nay**! (Nile, Amazon, Yangtze)

THE WORLD'S LARGEST DESERTS:
Deserts make me **sag**. (Sahara, Arabian, Gobi)

THE 7 CONTINENTS:
Eat an aspirin after a nasty sandwich. (Europe, Antarctica, Asia, Africa, Australia, North America, South America)

TRADITIONAL DIET FOR TREATING DIARRHEA:
BRAT (bananas, rice, apples, toast)

DIRECTIONS OF THE COMPASS, CLOCKWISE:
Never eat soggy wheat (north, east, south, west)

GUITAR STRINGS, FROM THICK TO THIN:
Elephants and donkeys grow big ears. (E, A, D, G, B, E)

COLORS OF THE RAINBOW:
Roy G. Biv (red, orange, yellow, green, blue, indigo, violet)

5 KNOWN DWARF PLANETS IN EARTH'S SOLAR SYSTEM:
Pluto can't make Eris Hot.
(Pluto, Ceres, Makemake, Eris, Haumea)

THE MOHS SCALE OF MINERAL HARDNESS, FROM SOFTEST TO HARDEST:
Toronto girls can flirt and only quit to chase dwarves.
(talc, gypsum, calcite, fluorite, apatite, orthoclase, quartz, topaz, corundum, diamond)

ORDERS OF COLOR ON A TV TEST PATTERN:
When you catch German measles
remain between blankets.
(white, yellow, cyan, green, magenta, red, blue, black)

FIVE FREAKY FACTS ABOUT...
INVENTIONS

- Seat belts were originally designed not for cars, but to secure workmen and window-washers to their equipment when scaling tall buildings.

- The first waterbed was developed more than 3,000 years ago, when Persians filled goat skins with water, sealed them with tar, and warmed them in the sun.

- After NASA developed a coating to protect its cameras from the sun's radiation, scientists at the California Institute of Technology used it to create radiation-blocking sunglasses. Most famous brand: Blu Blocker.

- When Frank Etscorn, a psychologist studying addiction, spilled liquid nicotine on his arm, he caught a buzz. He realized smokers trying to quit could be given nicotine through the skin. His nicotine patch hit the market in 1992.

- Early contacts were made from wax molds—by pouring wax over the eyes. The lenses were glass, and caused so much pain that patients were prescribed an anesthetic with cocaine.

NANO-GOLD

How tiny is a nanometer? It's one-billionth of a meter. Nanometers are the units of measurement for the world's smallest particles—atoms and molecules. Amazingly, scientists are now finding practical uses for particles of gold that measure in mere nanometers.

HOW TO CATCH GOLDFINGER

Despite advances in DNA evidence, forensic investigators still favor an old-fashioned method of crime-scene detection: fingerprints. These are obtained by applying chemicals that react with the amino acids in sweat that was left behind in the print. But prints last for only about three hours on nonporous surfaces, and people with very dry skin don't always leave clear fingerprints. Modern science has a solution: nano-gold. Researchers in Sydney, Australia, have found that mixing gold nanoparticles into those chemicals gives much sharper detail, no matter how old the prints are or what surface they're on. This is an important step toward the "holy grail" of forensic science: recovering prints from a crime victim's skin—even from corpses.

WASTE NOT

Thanks to nano-gold, sewage treatment plants could go from consuming energy to producing it. How? By using

the microbial fuel cell, a device that converts chemical energy to electrical energy. Bacteria from sewage are placed in the *anode chamber* of the cell, where they consume nutrients and grow, releasing electrons in the process. Result: electricity! Engineers at Oregon State University have discovered that coating the anodes with nano-gold increases the amount of energy released twentyfold. Sewage treatment plants might produce their own operating power, and even become "brown energy" generators.

DIVIDE AND CONQUER

More than a third of all Americans—about 120 million people—will be diagnosed with cancer sometime during their lives. When the wife of Dr. Mostafa El-Sayed, a Georgia Tech professor, was fighting breast cancer, he began looking into cancer research. "In cancer, a cell's nucleus divides much faster than normal," says El-Sayed. "If we can stop a cell from dividing, we can stop the cancer." El-Sayed felt that the properties of gold might be useful in killing cancer cells, and he designed nanometer-sized spheres of gold to test his theory. He and his team harvested cells from cancers of the ear, nose, and throat and coated them with a peptide that would carry the nano-gold into the cancer cells, but not into healthy cells. Result: The cancer cells started dividing, then collapsed and died. Though the discovery came too late to save El-Sayed's wife, nano-gold may save many lives in the future.

RESIGNED IN PROTEST

WHO: Bruce Boler, a water quality specialist with the Environmental Protection Agency (EPA)

BACKGROUND: Boler was assigned to southwest Florida in 2001 to assess the impact of development in and around the area's wetlands. In the course of his work he refused several permits for golf course developments because of the amount of pollutants the developments would discharge into sensitive wetland areas. Outraged developers funded their own "scientific" studies, which determined that developments such as golf courses were actually better for the environment...than natural wetlands. Amazingly, in 2003 the EPA accepted the studies.

RESIGNATION: Boler immediately resigned, calling the findings "absurd," and went public with the information. (He now works at Florida's Everglades National Park.)

IT'S SCIENCE!

Thoughts on the joys and frustrations
of scientific discovery.

"Men love to wonder, and that is the seed of science."
　　—*Ralph Waldo Emerson*

"The task is not to see what has never been seen before, but to think what has never been thought before about what you see every day."
　　—*Erwin Schrödinger*

"The whole of science consists of data that, at one time or another, were inexplicable."
　　—*Brendan O'Regan*

"A good scientist is a person in whom the childhood quality of perennial curiosity lingers on. Once he gets an answer, he has other questions."
　　—*Frederick Seitz*

"Science belongs to no one country."
　　—*Louis Pasteur*

"The most exciting phrase to hear in science, the one that heralds new discoveries, is not 'Eureka!' but 'That's funny...'"
　　—*Isaac Asimov*

HE DID IT HIMSELF

Medical breakthroughs sometimes begin with a visionary who asks, "What if..." and then proceeds to do something reckless enough to land him a write-up in *Ripley's Believe It or Not*. Werner Forssmann was just such a man. It was this 25-year-old German medical student who asked in 1929, "Can I thread a flexible tube all the way to my heart through a vein, and then photograph it for the rest of the guys to see down at the beer hall?" Without waiting for an answer, Forssmann cut an incision into the basilic vein in his upper arm. The plucky student then threaded a urethral catheter—a transparent tube used to help patients who can't urinate—into the vein. Werner walked down a flight of stairs with the tube in his arm, entered the radiology department, sat down on a table, and continued threading the tube toward his heart, using a mirror to watch his progress on a primitive X-ray device known as a fluoroscope. When he threaded the tube all the way into his heart's right atrium, Forssmann X-rayed the event for

posterity. At that moment, the diagnostic tool known as angiography was invented.

English physician William Harvey had proven in the 1600s that the blood circulates through the body and that the heart is the center of circulation. So, in theory at least, Forssmann had known it should be possible to measure the activity of the human heart by determining pressures inside the heart's chambers. If pressures were good, the heart was functioning normally; if low or erratic, the physician could diagnose faulty cardiac function.

A year later, Forssmann repeated his trick. Only this time, he injected an iodine compound through the hollow catheter and into the right atrium of his own heart for the world to see. So now not only the catheter but the area of the heart injected with iodine could be seen with an X-ray.

Because of Forssmann, the diagnostic tool known as cardiac angiography was invented...and patients everywhere could now get big fat bills for "diagnostic procedures."

CHEMICALS ARE COOL!

In 1997 high school students across America got time off from class to attend "Chem TV." Supposedly designed to get kids excited about chemistry and science, it was a traveling extravaganza featuring music, videos, lasers, games, skits, dancers, free T-shirts, a huge set with giant TV screens, and enthusiastic young performers.

Educational? Sort of. Chem TV (meant to sound like "MTV") said it was about *chemistry*, but it was really about the *chemical industry*. It was part of a million-dollar public relations campaign by Dow Chemical to help change its image. Dow had a controversial history: It supplied napalm and Agent Orange to the government during the Vietnam War, and lawsuits over faulty breast implants nearly bankrupted the company in 1995.

Critics charged that the Chem TV presentations were misleading (in one, an actor took off his clothes to demonstrate that "your entire body is made of chemicals"). Chem TV didn't differentiate between a *chemical* (a man-made combination of ingredients) and an *organic compound* (molecules that fuse together naturally—like water). Even so, the program toured for three years and won numerous awards. And it was tax exempt because it was "educational."

THE STINKY CHEESE STUDY

When most people think of a smelly cheese, Limburger is what comes to mind. Even though Mighty Mouse had a weakness for it, the stuff's indescribable scent repels most people. Researchers at the Netherlands' Wageningen Agricultural University found that the mosquito species *Anopheles gambiae* "loves both stinky feet and Limburger cheese." Limburger's uniquely smelly properties come from *B. linens*—the same bacteria that attract malaria-carrying mosquitoes to stinking feet. But despite that fact, Limburger is *not* the world's stinkiest cheese.

The *B. linens* bacterial strain is shared by many other washed-rind cheeses, including Muenster. During the ripening process for washed-rind cheeses, cheesemakers rinse the cheeses with a liquid (usually wine or beer), which encourages bacteria to grow and gives the cheeses a reddish-brown rind. Many people associate smelly with moldy, and moldy with old. But washed-rind cheeses are young cheeses, and the washing process is what brings this type of cheese to maturity. They're most common in Europe.

In 2004, scientists at Cranfield University in Bedfordshire, England, conducted a study to determine the most fetid fromage out of 15 contestants. The panel of pungency used 19 human testers backed up by an electronic "nose" to make its determinations. The electronic

nose was made up of an array of sensors, each of which responded to the presence of chemicals in a slightly different way. The sensors were linked to a computer that interpreted their responses and rated them.

OLFACTORY ASSAULT

The winner was Vieux Boulogne ("Old" or "Aged Boulogne"), a washed-rind cheese from France that testers likened to a combination of garlic, unwashed feet, and unwashed cat. But like Limburger, the actual taste of Vieux Boulogne (also known as Sable du Boulonnais) isn't as disgusting as its description. It's tangier and alcoholic because of the bière blonde (pale ale) used to wash its square rind. In fact, the owner of Le Fromagerie, the only store in London that stocks Vieux Boulogne, describes it as "a young, modern cheese with a surprisingly mellow and gentle taste that's perfect served with some crusty bread and a beer. It's a great cheese to try, as it doesn't have the earthy, farmyardy flavours that some people find overpowering."

Vieux Boulogne is made in the northernmost reaches of Normandy, a region of France renowned for its dairy products. The rich, flavorful milk from the cows near the city of Boulogne-sur-Mer contributes to its "mellow and gentle taste." But don't expect to sample it at your neighborhood grocery store. Vieux Boulogne is an unpasteurized milk cheese, so it's not legal in the United States.

The Tech Conference Hoax

BACKGROUND: Massachusetts Institute of Technology student Jeremy Stribling submitted an academic paper to a leading technology conference. The paper, entitled "Rooter: A Methodology for the Typical Unification of Access Points and Redundancy," was accepted, and he was invited to speak at the World Multiconference on Systemics, Cybernetics, and Informatics in Orlando, Florida.

EXPOSED! The paper was nothing but gibberish generated by a computer program. The program, written by Stribling and two fellow students, automatically spit out important-sounding nonsense, such as: "We can disconfirm that expert systems can be made amphibious" and "We concentrate our efforts on showing that the famous ubiquitous algorithm for the exploration of robots by Sato et al. runs in ▶((n + log n)) time [22]." Stribling later admitted to the hoax, adding that they'd done it because they were tired of being inundated with e-mail spam soliciting research papers for the conference. His conference credentials were subsequently revoked.

THE ICEMAN COMETH

On September 19, 1991, two German hikers spotted a body protruding from a glacier in the Ötzal Alps along the Austrian/Italian border.

- When scientists carbon-dated the remains, Ötzi the Iceman (as dubbed by the press) turned out to be more than 5,300 years old. Ötzi, a dark-skinned male around age 45, stood 5'5" tall and weighed about 134 pounds. He lived in the Copper Age around 3350 BC, and he continues to provide a treasure trove of information about Stone Age humans.

- Ötzi was the world's oldest fully preserved intact prehistoric human—a wet mummy complete with skin, hair, bones, and organs. Previous bodies had been dissected, embalmed, or destroyed by animals. Archaeologists were surprised by his shaved face and freshly cut hair.

- His body was found with everyday clothing and supplies—a copper ax, flint, dagger, bow, quiver, and arrows, which shed more light on daily life than the ceremonial objects typically uncovered with other bodies. (Unfortunately, visitors soon arrived, destroyed his clothing, and apparently stole his private parts.)

- Ötzi is the oldest tattooed human mummy yet discovered. So far 61 tattoos have been mapped on his body. They were made not with needles but with charcoal rubbed into cut marks. Their locations are consistent with acupuncture points, thus predating the first known use of this healing technique by 2,000 years.

- Theories about his death abound, but the latest forensic evidence suggests that Ötzi was murdered. He sustained defensive injuries in the weeks prior to his death that were consistent with a physical altercation. Pollen in his body shows that 24 hours before his death he climbed from the valley floor to the glacier where he died, perhaps to evade his adversary. The arrow found in his shoulder seems to have been launched from a great distance below Ötzi's location. The fact that the ax and other valuables were not stolen suggests that robbery was not the motive.

- DNA tests and body-scans reveal that he suffered from hardening of the arteries, gallstones, worn joints, parasites, gum disease, and that he was the first diagnosed victim of Lyme disease. Later DNA studies have revealed at least a dozen living genetic relatives in the Tyrol region. Ötzi's remains are stored in a special cold cell in Italy at the South Tyrol Museum of Archaeology, where scientists continue to study his life and cause of death.

"Scientific" Theory: Rain Follows the Plow

In the 1860s a theory became popular in the United States: It said that, when humans planted crops in dry, previously unsettled regions, moisture was released into the air, with the result being a change in climate and more rainfall. This became known as the "rain follows the plow" theory, and it was pushed by scientists, the local railroad company, and then the press, just as the government was trying to get people to settle the Great Plains—a region long known as a dry and barren wasteland. (The fact that the region happened to be experiencing a rare wet spell in the 1870s helped the theory take hold.) The theory comforted the millions of people who over the following decades moved to the Plains and built farms. No need to worry, promoters said—build it and the rain will come! It didn't. Millions of acres of farmland were subsequently wiped out during the repeated droughts that followed, culminating in the devastating Dust Bowl years of the 1930s.

According to the Latest Research...

- In a 2009 study by Oxford University, scientists gave a group of ducks full access to a pond, a water trough, and a shower. They found that the ducks preferred standing under the shower to standing in still water. The three-year study cost more than half a million dollars.
- In 2003 researchers at Plymouth University in England studied primate intelligence by giving macaque monkeys a computer. They reported that the monkeys attacked the machine, threw feces at it, and, contrary to their hopes, failed to produce a single word.
- In 2001 scientists at Cambridge University studied kinetic energy, centrifugal force, and the coefficient of friction...to determine the least messy way to eat spaghetti.
- Researchers at the Hungarian Academy of Sciences in Budapest analyzed videos of the "Wave" at sporting events. Results: it almost always moves clockwise around the stadium, travels at about 40 feet per second, and the average width of a wave is 15 rows of seats.
- A 2002 study by the Department of Veterans Affairs Medical Center in Vermont found that studies are often misleading.

ACCIDENTAL DISCOVERY: INSULIN

In 1889 Joseph von Mering and Oskar Minkowski, two German scientists, were studying the digestive system. As part of their experiments, they removed the pancreas from a living dog to see what role the organ plays in digestion.

The next day a laboratory assistant noticed an extraordinary number of flies buzzing around the dog's urine. Von Mering and Minkowski examined the urine to see why...and were surprised to discover that it contained a high concentration of sugar. This indicated that the pancreas plays a role in removing sugar from the bloodstream.

Von Mering and Minkowski were never able to isolate the chemical that produced this effect, but their discovery enabled Canadian researchers John J. R. MacLeod and Frederick Banting to develop insulin extracts from horse and pig pancreases and to pioneer their use as a treatment for diabetes in 1921.

Go to page 106 for the rest of the story.

IT REALLY IS A MELTING POT

In 2006 Professor Peter Fine at Florida Atlantic University asked his class to do a project on their own racial identities, and to submit to DNA tests as part of it. Twelve of the students considered themselves white with European ancestry; one considered himself black and of African descent. Results: Only one of the white students turned out to be completely European, and the black student turned out to have 21% European ancestry. The rest had various degrees of European, African, Native American, and East Asian genes. Professor Fine himself, who considered himself of typical white European stock, found out that he had 25% Native American genes. "I honestly think these tests could have a large effect on American consciousness," Fine told the U.K.'s *Observer* newspaper. "If Americans recognize themselves as a mixed group of people, that could really change things."

THE HAWKING WORMHOLE

"Wormholes are all around us, only they're too small to see," says physicist Stephen Hawking. He describes wormholes as "tiny crevices, wrinkles, and voids" in the universe that connect separate areas of space and time.

If scientists could devise a machine that would make a wormhole bigger—say, large enough for a human to pass through—it might make a handy time machine. Unfortunately, right now, Hawking doesn't know how to do that. What he does know is what he'd do with a wormhole time machine if he had one: "If I had a time machine, I'd visit Marilyn Monroe in her prime or drop in on Galileo as he turned his telescope to the heavens."

A vampire bat drinks more than its own weight in blood every night.

WEIRD ENERGY:
BODY HEAT

Cremation is an increasingly popular way of dealing with the dead—it's cheaper and more ecologically sound than traditional burials. In a typical crematorium, the furnace reaches a temperature of about 1,800°F. The deceased's body is reduced to ashes, but that's not the whole story. All the organic matter in the body, which is mostly water, vaporizes and becomes gaseous. Most nations require that the gases be treated before they're released. So the gases are passed through a cold-water chamber, which cools them to 350°F, and then a filter traps any gasified toxins—primarily the mercury in tooth fillings. The gas is released into the air, but the water in the cooling chamber is now very hot and nonpotable because it had cremation gases passed through it.

Solution: more than half of Denmark's crematoriums send the water into municipal heating systems. It was a controversial plan, but the Danish Council of Ethics (a board of scientists and clergy) approved it, as did the International Cremation Federation. The key was that the energy is being donated—if the water were being sold, it would violate a 1937 treaty that bars crematoriums from selling "residue of cremation." The plan went into effect in 2012, and the idea is spreading to other European nations. A low-income housing project in Paris, for example, is now heated with crematorium water, and a crematorium in England is using its wastewater to heat...itself.

HOW TO CREMATE A BODY

The modern process of cremation uses industrial ovens that can reach temperatures over 2,000°F. A regular oven isn't hot enough for cremation, but you can DIY with a homemade funeral pyre.

1] On first glance, a pyre may look like just a pile of wood, but it must be constructed a specific way to generate the heat necessary to turn your loved one into ash. Basically, you will be building a Jenga tower, so it's important that your slats are uniform or your pyre will be lopsided and Uncle Jerry could topple off midway through the burn. Start by placing a metal frame or grate on the ground for a base. This will get the wood up off the ground and allow airflow underneath.

2] Stack your wood on top of the metal grate in layers, each crisscrossing the layer below for stability. Also make each layer of wood slightly smaller than the one below, so the pyre narrows

as the stacks grow higher. This will create a chimney effect during the burn, concentrating the heat near the top to reach a maximum temperature of around 800°F. That's no industrial cremator, but it will get the job done.

3] Lay your loved one atop the pyre and light 'er up. Try to start the fire as centered as possible to encourage an even burn. (The metal frame also allows you to reach underneath the pyre.)

4] You can speed along the burn with an accelerant like kerosene, but if your loved one was an "all-natural" type, this may be against his or her wishes.

5] It will take around five hours for the body to be completely cremated. Body remnants from a pyre burn will be mostly indistinguishable from the wood ashes. So if you do want to collect and spread the ashes somewhere, plan to bring about five one-gallon buckets.

DUMB PREDICTIONS

"Space travel is bunk."

—SIR HAROLD SPENCER JONES,
ROYAL ASTRONOMER
OF ENGLAND, 1957
*(two weeks before the
launch of* Sputnik*)*

DARWIN'S MYSTERY MOTH

Our Tale of the Missing Proboscis is not a science-fiction-mystery film; it's the story of an animal whose existence Charles Darwin predicted in 1862. He had been studying the comet orchid of Madagascar, a "moth-loving" flower that is pollinated by moths when they feed on its nectar. The odd thing about this flower is that its nectary—the part that produces nectar—is more than 11 inches long, but the nectar is stored in its base. For a moth to feed on the nectar, it would need a proboscis (tongue) nearly a foot long. Such an insect had never been heard of, but Darwin claimed that if no such creature had evolved a proboscis that could access the nectar, the orchid would not be pollinated and would have gone extinct long ago. He published the theory in 1862...and it would take four decades for the theory to be proven correct.

In 1903 the Morgan's sphinx moth, with a tongue more than three times the length of its 3-inch-long body, was discovered in Madagascar. It's known today as the long-tongued night-flying hawk moth, and also the "predicted moth." In 2004 entomologist Phil de Vries went to Madagascar to film the moth feeding from the flower. Just finding the *flower* was hard enough, but de Vries did find it, and he set up a night-vision camera... and waited at the base of the tree, keeping his eye on the monitor. At 4:34 a.m., he noticed something—and watched in awe as a large moth flew up to the orchid, uncoiled a very thin, long proboscis, carefully inserted it into the blossom, and shoved it down the 11-inch nectary. The moth then sat there, flapping and drinking away. It was the first time that Darwin's predicted moth had been captured doing just what Darwin predicted it would do.

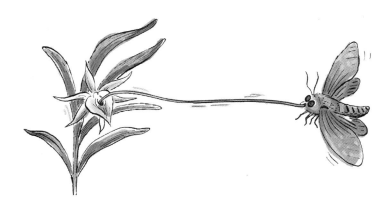

Unexplained Weight Gain

Sleep eating, a disorder that affects at least 1 percent of adults, is more than just a case of the munchies. Those who suffer from it have no conscious control over their actions while they raid the fridge. Patients may sleepwalk to the kitchen and prepare food while asleep, often leaving behind a mess as evidence. Dangers abound! One threat is to their waistline; sleep eaters tend to eat high-calorie and sugary foods such as peanut butter, chocolate, and even syrup straight out of the bottle. Their safety is endangered, too. They may cut themselves while chopping food; leave stove burners on; or choke while eating peanut butter by the spoonful, uncooked spaghetti, or frozen foods. Some eat inedible things like coffee grounds, eggshells, or even (in one case) buttered cigarettes. To top it off, medications prescribed for this condition may have pesky side effects, including...excessive drowsiness and restless sleep.

Q: What can you do that an astronaut can't?

A: Burp. (There's no gravity to separate the stomach gases from the liquids.)

MOVIE MAD SCIENTISTS

FRANKENSTEIN

There's the classic 1931 version starring Colin Clive as
Dr. Frankenstein and Boris Karloff as the grunting, rivet-
necked monster, or the not-so-classic 1994 version with
Kenneth Branagh as the good doctor and Robert De Niro as
the monster (which certainly puts a new spin on the classic
De Niro line, "You lookin' at *me?"*). The 1931 version is
indelibly printed onto our cultural memory—the collective
image of the Frankenstein monster is Boris Karloff's—but
on the other hand, the 1994 version is more faithful to the
original 19th century novel by Mary Shelley. And it's in
color! And don't forget *Young Frankenstein*, Mel Brooks's
dazzling send-up of *Frankenstein* and classic horror
films—a classic in its own right.

HOLLOW MAN

Kevin Bacon turns invisible, and no, this is not an
assessment of his movie career. In *Hollow Man* he plays an
unethical scientist who uses his own untested process to
become invisible. Then, as he must in a movie like this, he
goes completely insane and starts sneaking into hot girls'
apartments and killing off colleagues. Watch this for the
special effects; the story is a bit...transparent.

MUTTON AND A LITTLE LIVER

In March 2007, Professor Esmail Zanjani of the University of Nevada at Reno announced that he had successfully injected sheep fetuses with human stem cells. The result: sheep that grew organs that were part human. Some had livers, for example, that were made up of as much as 40% human liver cells. Zanjani hopes the research may one day lead to sheep being raised only for the human organs in their bodies—which could be transplanted into humans who need them. The scientists could conceivably create sheep that are tailor-made for specific people. For example, a sheep could be injected with your bone marrow in order to grow organs suitable just for you. Zanjani insists that the work is ethical and medically necessary, and that the sheep are not monsters. "We haven't seen them act as anything but sheep," he says.

WHISKER SCIENCE, Part I

*A cat's whiskers are a marvel
of form and function.*

- The scientific name for whiskers is *vibrissae*, and they're specialized sensory organs on a cat's body (mostly on the cheeks). On average, cats have 24 cheek whiskers.

- Each whisker is double the thickness of an ordinary hair and is rooted in the cat's upper lip. Every root connects to 200 or more nerve endings that transmit information directly to the cat's brain.

- Air currents create a tiny breeze as they move around an object. Cats feel this change with their whiskers and avoid objects in their path.

- Whiskers direct hunting cats to a successful pounce. In one experiment, a blindfolded cat was placed in an enclosure with a live mouse. When the cat's whiskers touched the mouse, the cat grabbed his prey and delivered a killing bite in one-tenth of a second.

- Once the prey is in the cat's mouth, facial muscles allow the whiskers to curl forward and sense any movement that might mean the animal is still alive and possibly dangerous.

NO TIME LIKE THE PRESENT

Real ideas about time travel start with Albert Einstein's theory of special relativity. The theory says that the faster you travel through space, the slower you travel through time. If you hop a plane from New York to London, traveling at, say, 500 mph, by the time you get home, you'd be about 30 nanoseconds younger than the friend you left behind. Neither of you would notice, of course. But if you could travel through space at the speed of light for 10 years, by the time you returned, hundreds of years would have passed, and everyone you knew would be long dead. Your "present" and your friend's "present" would now be...1,000 years apart.

Einstein's second theory—on general relativity—says that space-time can be bent. One point in time can be bent to touch any other long-past or far-future moment. Think about placing a heavy object on an outstretched piece of elasticized fabric. The fabric sags until it eventually curls in on itself and its ends touch. If the past can touch the present, is it still the past or is it now the present?

Tesla's Theory of Relativity

Did Nikola Tesla (1856–1943) come up with a better theory of relativity than Einstein? He argued that Einstein's idea of curved space is bunk:

> I hold that space cannot be curved, for the simple reason that it can have no properties. It might as well be said that God has properties. He has not, but only attributes and these are of our own making. Of properties we can only speak when dealing with matter filling the space. To say that in the presence of large bodies space becomes curved is equivalent to stating that something can act upon nothing. I, for one, refuse to subscribe to such a view.

So where is Tesla's theory of relativity? His work is shrouded in mystery, and conspiracy theories abound. He *did* say that he came up with one that would "put an end to idle speculations and false conceptions, as that of curved space," but no one has been able to find it in any of his writings. Or...it's been classified.

BLOODSTREAM

- A pumping human heart can squirt blood as far as 30 feet.

- You can lose up to a third of your blood and still survive.

- The human body has about 60,000 miles of blood vessels.

- In the time it takes to turn a page, you'll lose 3 million blood cells and make 3 million more.

- Red blood cells live four months. In that time they make 75,000 trips to the lungs and back.

- The most nutritious "food" in the world is blood.

- The Rh factor in blood occurs much more frequently (40–45 percent) in Europeans and people of largely European ancestry.

- Blood is thicker than water: blood has a specific gravity of 1.06; water's is 1.00.

- Identical twins always have the same blood type.

- The average number of industrial compounds and pollutants found in an American's blood and urine: 91.

SURFING SCIENCE

Surfing is hundreds of years old. Ancient Hawaiians surfed on big, heavy, wooden boards. In the 1950s the sport caught on in Southern California, where new engineering techniques and materials like fiberglass allowed for lightweight, smaller boards that still supported the mass of the surfer on the water. Spectators flocked to the amazing sight of swimmers standing on seemingly flimsy surfboards while cresting the tops of breaking waves. Most did not know they were watching some excellent feats of rotational motion and physics.

Balance is obviously an important part of surfing. How does a surfer stay stationary (balanced) on a board that's cresting a wave? Along the lengthwise center of the board and slightly toward its tail, where there's extra mass, lies the center of the board's gravity. This point is the board's axis—like the fulcrum at the center of a seesaw. Where the surfer stands in relation to the axis controls his or her board's rotational motion exactly like the up-and-down rotational motion of a seesaw. If the rider's weight moves too far toward the nose of the board, the board tips (or torques) forward and the nose sinks. Too far back and the tail sinks. A good surfer straddles the center of gravity with one foot toward the tail and one toward the nose. The two torques cancel each other out, and the surfer is balanced.

But it takes more than an understanding of rotational motion to make a brilliant surfer. Our surfer needs a thorough (even if intuitive) understanding of the development of potential energy and how it can be turned into kinetic energy. A surfer arrives at the top of a wave just before it breaks. By taking up this position, she has gained potential energy. Potential energy is the potential product of you and your equipment's weight or mass, and the vertical distance you're about to fall. Our surfer converts this potential energy to kinetic energy when she drops off the top of the wave down toward the flat of the wave. This conversion into energy gives her the power to propel herself along despite the friction of the water currents. The surfer can now ride the wave.

In 1929 the Nazis drew up plans to put a giant mirror in space to reflect the sun's light as a weapon. If they had succeeded, their "Sun Gun" could have destroyed entire cities.

The Smartypants Family

Marie Curie was the first female scientist to be awarded the Nobel Prize. She actually won it twice, first in 1903 for Physics, and then in 1911 for Chemistry (for her discovery of radium and polonium), making her not just the only woman to win two Nobels, but the only scientist—male or female—to win it in two different sciences. It turns out Curie's scientific prowess is a family trait:

- **Marie's husband, Pierre Curie**, was also awarded the 1903 Nobel prize for their combined work in using radioactive isotopes to treat tumors.

- **Marie's daughter, Irène Joliot-Curie**, won the 1935 Nobel Prize for Chemistry (along with her husband) for discovering artificial radioactivity.

- **Marie's grandaughter, Hélène Langevin-Joliot**, is a famous nuclear physicist.

- **Marie's grandson, Pierre Joliot-Curie**, is a well-known biologist.

- **Marie's great-grandson, Yves Langevin**, is an astrophysicist.

In all, the Curie family has won a record five Nobel Prizes...so far.

ROBOTS IN THE NEWS

THINK-BOT Computer scientists at the University of Washington have figured out a way to control a robot's actions. What's so unique about this new programming technique? The scientists control the robots *with their minds*. They gave a humanoid robot a special cap outfitted with 32 electrodes. On the other end of the electrode cap was a human subject in a similar cap who, just by thinking about telling the robot to move forward or grab an object, was able to do so via the power of brain waves.

VOTE-BOT In 2006 the British government released a report called *Robo-rights: Utopian Dream or Rise of the Machines?* It outlined possible major social problems that could result in England one day should robots develop artificial intelligence and become independent. The report says the government may have to provide robots with housing, health care, and the right to vote. Robots, meanwhile, would be expected to pay taxes and serve in the military.

IT'S NOT ROCKET SCIENCE

"It's not rocket science" is a vaguely insulting phrase people say to somebody struggling to solve a problem. It means "it's not that hard," bringing up a mock parallel with rocket science, which involves difficult math, chemistry, and mechanical engineering.

The phrase entered the lexicon in the 1940s. That was about when rocket science, or the world of aerospace, began in earnest. Some of the German scientists responsible for making the Nazis' V-2 rockets, including Wernher von Braun, surrendered to Allied troops in 1945. They were then recruited to work for the United States, and the world of rockets kicked into high gear. It was especially true after the end of World War II, when rocket science could be used for space exploration rather than weaponry. By 1950 rocket science, considered science fiction only a decade earlier, was very real but still so exotic that it was synonymous with "incredibly difficult." It was a common slang term, but print references didn't appear until the 1980s.

Before these brave new frontiers of science were crossed, however, how were people dismissive of others trying to do something hard? They used phrases like "as easy as pie"or "as easy as falling off a log."

A HUMOUROUS STORY

Doctors in ancient Greece and Rome believed the human body is filled with various fluids, which they called "humours." They identified four fundamental humours: black bile, blood, yellow bile, and phlegm. The human body is entirely composed of these four substances, they decided, and good health depended on their being in proper balance.

Each humour became associated with a particular organ (gallbladder, liver, spleen, or the brain and lungs, respectively), then with one of the four classical elements of earth, air, fire, and water. Each element, and each humour, was defined as either cold or warm, and either moist or dry. Blood, for instance, was identified with air, with warm and moist qualities, while black bile was associated with earth, cold and dry. The system encompassed the entire physical world. Each humour was identified with one of the seasons, with a particular time of day, with certain plants, animals, and minerals. Every planet, constellation, and astrological sign had a corresponding humour. Even human psychology was explained by four basic personality types—melancholy, sanguine, choleric, and phlegmatic—all defined by a dominant humour.

It made for bad medicine. In theory, eating the proper foods would encourage production of a particular humour—but rather than topping off one humour, medieval physicians preferred to correct imbalances by draining any excess of the others. If there wasn't enough blood in your mix, for instance, enemas would be prescribed to reduce your black bile, along with induced vomiting to bring up yellow bile. A rubdown with blistering agents would release phlegm in the form of weeping pus. If too much blood was your problem, there was bloodletting, sometimes accomplished by leeches. If treatments were ever effective, it was pretty much by accident. Why did this system remain popular for hundreds of years, until it was finally displaced by the scientific method? Perhaps because it gave a framework of order to a chaotic universe and placed humankind at the center of it. This theory was very comforting to the medieval mind—even if it wasn't true.

3 FACTS ABOUT BEES

1. Bees have different dialects. A German bee cannot understand an Italian bee.
2. Honey never spoils.
3. Bees use ultraviolet vision to see which flowers have the most nectar.

MAGGOT THERAPY

It sounds like something from a horror film—fat, cream-colored maggots eating their way through infected sores and wounds. It's not. It's medicine. Since ancient times, doctors have used maggots to prevent wounds from getting infected. In the 1940s, antibiotics replaced maggots. But bacteria adapted and started to become resistant to antibiotics.

RETURN OF THE MAGGOTS: Maggots work by secreting digestive enzymes that feed on dead tissue. Those enzymes also kill bacteria in a wound and speed up healing. Doctors place between 200 to 300 maggots on a wound, then cover it—maggots and all—with mesh. Beneath the mesh, the maggots feed for 48 to 72 hours. When they're done, the doctors remove them. Wounds that haven't healed for months, even years, often respond quickly to maggot medicine.

3-D Printing a New You

A printer that can replicate the human body? Sure, why not. One day, advanced 3-D scanners will scan you, and then organic ink and special plastics will "bioprint" made-to-order body parts. So far, skulls, eyes, skin, noses, ears, bones, and limbs have all been reproduced by 3-D printers. A 22-year-old woman in the Netherlands suffered from a condition that thickened the bone structure of her skull, causing headaches and loss of vision, and threatened to impair brain function. So in 2014, doctors were able to duplicate and replace her damaged skull with a 3-D printed plastic version. According to the lead doctor, brain surgeon Dr. Bon Verweij, it was a resounding success. The patient regained her vision, was back to work, and "there are almost no traces that she had any surgery at all."

DUMB PREDICTIONS

"Everyone's always asking me when Apple will come out with a cell phone. My answer is, probably never."

—TECH COLUMNIST DAVID POGUE, 2006

THE MONSTER STUDY

In 1939 Wendell Johnson, a speech pathologist at the University of Iowa, directed graduate student Mary Tudor in an experiment on six children, aged 5 to 15. For six months Tudor visited the kids (who lived in a nearby orphanage) and conducted classes with them—and whenever they spoke she told them they had terrible speaking voices. She did this again and again. Why? To see if she could turn the kids—all of whom had perfectly normal speaking voices—into stutterers, thereby proving Johnson's hypothesis that stuttering is caused by conditioning rather than congenital defect. None of the kids became stutterers, but according to Tudor's own notes, all of them became afraid and ashamed to speak. When the experiment was over, Tudor simply left. The children were never told anything. Johnson's peers were aghast when they heard what he'd done and dubbed his work "The Monster Study."

UPDATE: The results were concealed for decades, but in 2001 they were discovered by a journalist. Later that year the University of Iowa issued a formal apology for the experiment. In 2007 three of the test subjects, along with the estates of the other three, were awarded $925,000 by the State of Iowa for what all described as the lifelong scars they suffered as a result of the "Monster Study."

THE PERFECT FIRESTORM

A fire tornado is a rare occurrence that results when a fire is whipped into a burning frenzy by intense winds. Here's how this phenomenon works: First, a strong updraft of hot air hits a wildfire. As the hot air rises, it makes room for outside air to flow in. As that air whips in, it can form a whirlwind that picks up the flames and becomes a swirling column of fire nicknamed a "fire devil" or a "whirl." Fire devils often range from 30 to 200 feet high and usually last a few minutes. Like regular tornadoes, though, they can reach more than half a mile high with winds of 100 miles per hour...and they can be just as deadly. The worst fire devil occurred after the 1923 Great Kanto earthquake in Tokyo, when 38,000 people were killed in 15 minutes.

TREK*NOLOGY

Star Trek premiered on TV in 1966. Though we thought it would be centuries before humans developed technology to match *Star Trek*'s, some innovative thinkers are already turning "*Trek**nology" into everyday *tech*nology.

PHASER

Trek*nology: When exploring an alien planet, the crew of the *Enterprise* had to be ready for anything. Their best defense? The phaser—a handheld ray gun. Set on "stun," a phaser would merely immobilize the enemy; set on "maximum," it would vaporize him.

Technology: We don't have phasers yet, but several companies are trying to design them. Though the *Star Trek* phasers fit in the palm of your hand, most prototypes are much larger—one is parked on top of a 20-foot shipping container! The energy sources being explored range from lasers to microwave radiation, and uses include stunning (or immobilizing) an opponent or even frying the electronic components of a drone. When this technology is perfected, it will be reserved for military and police use.

COMMUNICATORS

Trek*nology: These small portable communication devices could be used anywhere, anytime, and also worked for remote tracking and locating. No operators. No phone booth. No cord. Sweet!

Technology: The first cellular phone call was made in April 1973, but it wasn't until 1983 that cell phones became available to the public—17 years after the first *Star Trek* episode. Today, cell phones not only act as communications devices, they can also log on to the Internet or offer navigation. And they are even smaller than the communicators Spock and Kirk used.

TRANSPORTER

Trek*nology: "Beam me up, Scotty!" Within seconds, Captain Kirk and his landing party would vanish from a planet's surface and reappear in the transporter room on the *Enterprise*. Teleportation was a means of transporting people from one place to another by converting them into pure energy, then changing them back into people again at the other end.

Technology: Scientists haven't been able to teleport a person (or even an object) from one place to another, but in 1998 they did succeed in teleporting a laser beam. When it's perfected, this technology will most likely be used for moving information—called quantum computing—and will allow people to move huge blocks of digital data at the speed of light. No more twiddling your thumbs while you download a game. But you'll have to wait a little longer before you can say "Beam me up, Scotty."

The Great Shakespeare Hoax

The "Flower Portrait" is probably the best-known
painting of William Shakespeare. The familiar portrait,
showing the Bard wearing a wide white collar pressed
tight up to his chin, has been reproduced countless
times. (It is often printed on the cover of programs
for Shakespeare plays.) It was named for one of its
owners, Sir Desmond Flower, who donated it to the
Royal Shakespeare Company in 1911. According to the
date on the reverse side of the picture, it was painted
in 1609—while Shakespeare was still alive.

Here's how science exposed the truth: In
2004 experts at London's National Portrait Gallery
conducted a four-month study of the painting,
using X-rays, ultraviolet light, paint sampling, and
microphotography. Their conclusion: It's a fake. It
was painted between 1814 and 1840, 200 years *after*
Shakespeare's death. They have no idea who painted it.

IT'S SEXY TIME

Science and the news intersex...er, intersect.

A WORLD OF LOVERS

Are you one to kiss and tell? If you are, go to
JustMadeLove.com—and tell the world about it. The
site, which uses the Google Maps program, allows
you to zoom in to your exact location and enter your
information, and it shows up as a marker on a map
on the JustMadeLove.com site. You can even leave
a comment about how "it" went. A sample, from
Greenland: *"böyle bi̦si yokk doymuyor istıyor en son
kanattımm onu pes etti yarım saat sonar bı daha ıstıyor
hep ısıyor."* (We hope that's not dirty.)

LOVE–BOT

David Levy, an expert on artificial intelligence, theorizes
in his book *Love and Sex with Robots: The Evolution
of Human-Robot Relationships* that by the year 2025
humans will be able to engage in realistic romantic
situations with robots. Levy believes that intimate
relations with ultrarealistic, humanoid-looking robots
will be commonplace by then, perfect for anyone who
might have difficulties attracting a mate.

Lighting Up the Dark

How many galaxies have you examined? Probably not as many as Vera Rubin, who studied hundreds of them. Of course, in the late 1940s, when she tried to enroll in the graduate program at Princeton (after being the only astronomy major at Vassar Women's College), she was told that women weren't allowed, so she "settled" for Cornell.

Then she made one of the most groundbreaking discoveries in the history of astronomy. While working with a large telescope in the 1970s, Rubin noticed that galaxies rotated much faster than previous astronomers had predicted they would. Newtonian physics said they should

"fly apart," but something was holding them together. "This unexpected result," wrote Rubin, "indicates that the falloff in luminous mass with distance from the center is balanced by an increase in nonluminous mass." Another way to describe "nonluminous mass": dark matter. Rubin had provided the first real evidence that dark matter exists—a mystery since Dutch astronomer Jacobus Kapteyn coined the term in 1922. After Rubin died in 2016, the *New York Times* described her contribution as "a Copernican-scale change in cosmic consciousness."

Would you pay to watch married couples bicker?

A study conducted by the National Institutes of Health observed 82 married couples to determine what factors make them happy. The conclusion? "The marriages that were the happiest were the ones in which the wives were able to calm down quickly during marital conflict."

Cost to taxpayers: $325,525

MYTHUNDERSTANDINGS

HIPPOCRATES

MYTH: Hippocrates was the father of modern medicine.

THE TRUTH: Thanks to the Hippocratic oath, which is still administered during the graduation ceremonies of many medical schools, the name of the ancient Greek physician has become virtually synonymous with the practice of medicine. He may have tried to heal people, but like all doctors of his era, Hippocrates knew virtually nothing about the workings of the human body. And almost all of what he did believe—for example "that veins carried air, not blood, and illness was caused by vapor secreted by undigested food from unsuitable diets"—was dead wrong.

CYBER COWBOYS

The term "hacker" dates back to 1960, when students at the Massachusetts Institute of Technology spent long hours "hacking" away at their keyboards in their Artificial Intelligence class, trying to make the computer do something it hadn't been programmed to do. Over the years, as hackers discovered they could break into all sorts of systems, they had a choice to make: Should they hack for the good of others...or for themselves? Like cowboys in Western movies, they had to choose whether they'd wear the black hat or the white hat.

BLACK HATS are the "bad guys" who break into corporate computer systems, stealing credit card numbers, bank accounts, identities, and e-mail addresses. They either use them for their own benefit or to sell or trade to other Black Hat hackers. Because of the criminal nature of their activities, the ethical hackers often call them *crackers*.

WHITE HATS are the "good guys"—security experts hired to protect companies from the Black Hats. White Hat teams find a hole in a company's security system and show the company how to fix it. In the White Hat community, there's an ethical code: They are loyal to their employers and sneer at the greed, theft, and vandalism of Black Hats.

BLUE HATS are an offshoot of the White Hats. They're ethical hackers, but they operate outside of computer security firms and are often contracted to test a system for bugs before it launches. The concept was created by Microsoft to find vulnerabilities in Windows.

GRAY HATS follow their own code of ethics; they are a little bit white and a little bit black. They don't actually *steal* assets—they find a hole or a bug in a company's security system (through illegal means), but often report their findings to the company and offer to fix it...for a hefty fee.

• • • • • • • • • • • • • • • • • • • •

PATENTLY WEIRD:
Initiation Apparatus, 1906

Before fraternities relied heavily on alcohol to enhance the initiation process, this electric shock treatment helped spark the fun. The apparatus, described as "entirely harmless in its action and results," was specifically "designed for use in lodges and secret societies." Two metal rails about an inch wide are laid down as tracks and hooked up to a battery or generator. The victim—pledge or inductee—then walks down the track wearing a pair of shoes with metallic soles, heels, and interior contact plates. Every time the subject takes a step, the electric circuit is opened and closed, continuously shocking whoever dons the metal slippers.

STONE MAN SYNDROME

Technically known as *fibrodysplasia ossificans progressiva*, this is a particularly perverse disease because it's the result of a malfunction of the body's own repair system. When nonaffected people injure fibrous tissue—things like muscles and tendons—their bodies jump into gear to fix the problem. But in the case of people with this disease, their bodies go too far and cause the injured tissue to *ossify*...or turn into bone. Real, actual bone. That means if you injure your wrist, your body responds by turning all the tendons, ligaments, and muscles in your wrist to bone, with the obvious result that you can't move your wrist anymore. In the most extreme cases, the victim can be rendered completely immobile, hence the "stone man" name. There is no cure, and there's not even any effective treatment; surgery to fix the ossified tissue just results in the body rushing to add more bone to the area.

THE GREAT GPS TREASURE HUNT

Have you ever wanted to go on a treasure hunt? Does trying to reach a random location sound like a great way to spend a Saturday afternoon? Then you may want to try geocaching. The concept is simple. After jotting down coordinates via a website or logging onto an app with their smartphones, geocachers attempt to reach a certain location and uncover a logbook. If they're successful, they add their name to it and confirm online that they found it.

To make finding a geocache more exciting, participants also sometimes add "trackables" to the stash: small objects like dog tags or coins that can be tracked as they make their way around a country.

A geocache is typically placed in a public park, but many are located in libraries, underwater, and even on mountaintops. Finding the trickier ones can involve answering riddles or visiting multiple locations before determining the correct spot. Tracking down other challenging geocaches often requires climbing trees or digging them up with a shovel.

But all that searching around can look...odd. Many participants have been questioned by authorities for "behaving suspiciously." Several geocachers have also died while searching, and a family in Rochester, New York, had to be rescued after looking for a geocache in a cave in 2012. As a result of these incidents, several government agencies have cracked down on geocaching.

"Scientific" Theory:
ICE MOON

Hanns Hörbiger was a successful and wealthy Austrian engineer who had lots of time to gaze at the sky through his telescope and develop weird theories about things. In 1913 the weirdness peaked when he published *Welteislehre*, or the "cosmic ice theory." Hörbiger explained that while staring at the moon one night, he theorized it was made of ice. But wait, there's more! That was followed by a vision in which Hörbiger learned great truths regarding the formation of the universe...and it all had to do with ice. Hörbiger promoted his theory relentlessly, and even though most serious scientists dismissed it, by the late 1920s the cosmic ice theory was very popular in Germany— largely because many Nazis had taken it up. (Heinrich Himmler, one of Hitler's commanders, became a major supporter of the theory, based on the idea that the "ice" somehow related to the Nazi fantasy of the "white Nordic race," which purportedly originated in icy regions of northern Europe.) Hörbiger died in 1931, but his theory remained popular and retained supporters right through World War II...until Germany's defeat and the war's end, when it quietly melted away.

HANGOVER SCIENCE

What causes a hangover? Read on.

- Your liver processes alcohol into a toxic chemical called *acetaldehyde*. Just as the alcohol made you feel good (or at least drunk), the acetaldehyde makes you feel bad. It's the accumulation of this chemical in your body, more than the alcohol itself, that causes hangover symptoms. (That's why the hangover comes *after* you've been drinking—the alcohol has been changed into acetaldehyde.) Specifically, acetaldehyde causes your blood vessels to dilate, which makes you feel warm, and can give you a headache.

- Meanwhile, the alcohol that's still in your system is raising both your pulse and blood pressure, which makes the headache even worse.

- And then there's the effect on your kidneys. When you're sober, your kidneys use a chemical called vasopressin to recycle the water in your body. But alcohol reduces the level of vasopressin in your body—which, in turn, reduces your kidneys' ability to function. So instead of recycling water, you urinate it out. That makes you dehydrated, which can make your hangover worse.

- It's also possible that what you're experiencing in a hangover is a minor case of alcohol withdrawal syndrome—the same thing that chronic alcoholics experience when they stop drinking. "Your brain becomes somewhat tolerant over the course of an evening of heavy drinking," says Dr. Anne Geller, who runs the Smithers Alcoholism Treatment Center in New York City. "The next morning, as the alcohol is coming out of your system, you experience a 'rebound.' You might feel nauseous, maybe you'll have some diarrhea, maybe you'll feel a little flushed. Your tongue is dry, your head is aching and you're feeling a little bit anxious or jittery."

PREVENTIVE MAINTENANCE

There are a few things you can do *before* you start drinking that may prevent the worst excesses of a hangover:

• Eat a substantial meal or at least have a glass of milk before you start drinking. It will help protect your stomach lining.

• Avoid champagne and dark-colored drinks, especially red wines. They contain byproducts of fermentation that may make the hangover worse.

• Drink a pint of water before you go to bed. The water will help minimize dehydration.

• The next morning, eat eggs. Studies suggest that eggs help stabilize blood sugar levels and replenish depleted B vitamins. A protein found in eggs, cysteine, is believed to help break down the acetaldehyde that's causing the hangover.

MEAD'S CREED

Margaret Mead, perhaps the most famous anthropologist in the world, helped shape our understanding of human behavior.

"The solution to adult problems tomorrow depends on large measure upon how our children grow up today."

"No matter how many communes anybody invents, the family always creeps back."

"The only thing worth doing was to add to the sum of accurate information in the world."

"Every time we liberate a woman, we liberate a man."

"What people say, what people do, and what they say they do are entirely different things."

"Our humanity rests upon a series of learned behaviors, woven together into patterns that are infinitely fragile and never directly inherited."

"One of the oldest human needs is having someone to wonder where you are when you don't come home at night."

"Nobody has ever before asked the nuclear family to live all by itself in a box the way we do. With no relatives, no support, we've put ourselves in an impossible situation."

"Sister is probably the most competitive relationship within the family, but once the sisters are grown, it becomes the strongest relationship."

"Thanks to television, for the first time the young are seeing history made before it is censored by their elders."

"Always remember that you are absolutely unique. Just like everyone else."

Charles Osborne of Anthon, Iowa, holds the title of "World's Longest Hiccuper." It started in 1922, hiccuping as often as 40 times per minute. Sometimes he hiccuped so hard his false teeth fell out. In 1987— nearly 70 years later—the hiccups stopped.

FIVE FREAKY FACTS ABOUT...
MINERALS

- Gold is so rare that all of the pure gold produced in the last 500 years would fit inside a 60-foot-square cube.

- Most minerals are luminescent. When they're exposed to radiation (like UV light from sunlight), they absorb and store it. When the mineral is then heated to at least 500°F, such as in a kiln, it releases that stored energy as light, and it glows.

- The diamond is the hardest natural mineral—four times harder than the next hardest, sapphire and ruby.

- When a body (human or otherwise) is buried, its bones absorb minerals, like fluorine, from groundwater. A technique called fluorine dating is used to approximate the age of skeletal remains.

- The element uranium was first uncovered in 1798 from inside a mineral called pitchblende.

QUANTUM MECHANICS 101

What's a book about strange science if we didn't attempt to explain quantum theory in two short pages?

The name sounds intimidating, but it can be broken down like this: A quantum is the tiniest bit of energy inside an atom. And mechanics is the study of motion. So quantum mechanics is the study of the motion of particles inside atoms. An electron, for example, is one such particle that orbits an atom's nucleus. But it doesn't move like a planet orbiting the Sun. Experiments have shown electrons can behave like waves. Trying to predict where one will be at any moment is a guessing game because electrons can instantly jump from one orbit to another—making quantum leaps.

It gets weirder. Two quanta (the plural of quantum) can share a mysterious link, even if they're far apart. In one experiment in 2012, scientists took a connected pair of photons (quanta of light) and separated them. Photon 1 was then altered, and amazingly, Photon 2, on an island 88

miles away (143 km), was instantly altered as well. The two were entangled, a strange connection between quanta that no one can explain.

Scientists have managed to create useful inventions using quantum mechanics—one of the first was the transistor, in 1947. This world-changing device replaced vacuum tubes, which were big, breakable, and needed a lot of electricity. Transistors made the computers of today possible, as well as digital cameras, CD and DVD players, cell phones, ATMs, and lasers. Still, there's much about quantum mechanics that scientists can't explain. Even Richard Feynman, a quantum physicist, once said, "I think it is safe to say that no one understands quantum mechanics."

ALL IN GOOD FUN
Nikola Tesla was close friends with Samuel Clemens, better known as Mark Twain. When Tesla made an X-ray gun, he shot it at Twain's head to capture several images of his skull on film.

Heart to Heart

In 1996 renowned English heart surgeon Sir Magdi Yacoub operated on two-year-old Hannah Clark of Mountain Ash, Wales. She had cardiomyopathy, which had caused her heart to become badly inflamed, so Yacoub put in a new one. Hannah and her new heart were doing well...until the heart started showing signs of rejection 10 years later. In 2006 Yacoub came out of retirement and operated on Hannah, now 12, again. He reconnected her original heart...which had been left inside her body

(surprisingly, not an unusual practice). It had apparently healed itself while it wasn't being used and began working immediately when it was reattached. Hannah was back home within five days. It was the first time that such an operation had ever been performed.

VIRGIN BIRTH

*Some insects and other invertebrates don't
need to mate to reproduce, but for vertebrates,
reproducing that way is rare—less than 0.1
percent have done it. Here is one.*

WHO? A boa constrictor snake

WHERE? At a pet store in Tennessee

WHAT? Although she had given birth naturally years
before, this female boa delivered twice more in 2009
and 2010 under different circumstances. DNA testing
determined that none of the male snakes she lived with
were the father of her 22 (only female) babies. It turned out
that all of them were "half-clones" (two eggs fused together)
of their mama and had no dad. This marked the first time a
boa had ever reproduced asexually.

WHY? No one seems to know. Interestingly, the boa
mom did apparently need to be around males before she
could become pregnant, though, even if she didn't breed
with them. Geneticist Warren Booth points out, "Only in
years that she was housed with males has she produced
offspring. It appears that some interaction with a male is
required. However, why she does not utilize his sperm is at
present unknown."

DREAM DISCOVERY: INSULIN

On occasion, scientific discoveries and inventions have resulted directly from a dream. Here is one.

Frederick Banting, a Canadian doctor, had been doing research into the cause of diabetes, but had not come close to a cure. One night in 1920, he had a strange dream. When he awoke, he quickly wrote down a few words that he remembered: "Tie up the duct of the pancreas of a dog...wait for the glands to shrivel up...then cut it out, wash it... and filter the precipitation." This new approach to extracting the substance led to the isolation of the hormone now known as insulin, which has saved the lives of millions of diabetics (sadly, at the cost of some dogs' lives). Banting was knighted for his discovery.

HOLLYWOOD PHYSICS

You've seen these things happen in the movies, but can you tell the science from the fiction?

FICTION: After falling off a cliff, a car will burst into flames on impact.

SCIENCE: A car will only burst into flames if the gas tank is severely damaged, the gasoline has been vaporized, and there is a source of ignition...like a lit match.

FICTION: Stray bullets give off a spark or a flash of light when they strike a surface.

SCIENCE: Most handgun bullets are made of copper-lead or lead alloys, which don't spark when they strike a surface, even if the surface is made of steel.

FICTION: Automatic weapons can fire off thousands of rounds for several minutes at a time without being reloaded.

SCIENCE: It's true that automatic weapons can fire off thousands of bullets—but only for a few seconds at a time. Sustaining a ten-minute gun battle would not only require tens of thousands of bullets, it would also overheat the guns and cause them to malfunction.

How Color Vision Works

The cones in our eyes allow us to see in color.
Here's how color vision works:

- Seeing means seeing light that is reflected off objects. That light enters our eyeballs, goes through the lenses, and hits the retinas. There it affects specialized cells called photoreceptors, which send neural signals to the brain's visual center.

- There are two main types of photoreceptors in the retina: rods and cones (so called because of their shape). Rods detect different amounts of light (bright to dark), and cones detect different wavelengths of light—meaning different colors.

- Humans have three types of cone cells, which give us trichromatic (three-color) vision. One type of cone cell responds to short-wavelength light (the blue part of the spectrum), another to medium-wavelength light (the green part), and the third to long-wavelength light (the reds).

DUMB PREDICTIONS

"Television won't be able to hold on to any market it captures after the first six months. People will soon get tired of staring at a plywood box every night."

—Film producer Darryl Zanuck, 1946

OLD HISTORY, NEW THEORY

THE EVENT: The Black Death, which wiped out about a third of the population of Europe in the mid-1300s

WHAT THE HISTORY BOOKS SAY: The Black Death was caused by an outbreak of bubonic plague, spread by rats.

NEW THEORY: Dr. James Wood, professor of anthropology at Penn State University, used computer analysis of church records and other documents to map out how the plague spread across Europe. If the epidemic really had been caused by bubonic plague, it would have spread differently than this one did.

Rat-borne bubonic plague has to reach epidemic levels in the rat population before it can cause an epidemic in humans, Wood says. The Black Death seems to have spread faster among humans than it could possibly have spread among rats. And there's little evidence of a rat epidemic in the historical record. "There are no reports of dead rats in the streets in the 1300s," he says.

Also, symptoms of bubonic plague are stark and unmistakable: high fever, bad breath, body odor, coughing, and vomiting of blood, followed by swollen lymph nodes and red bruising on the skin that turns purple and then black. Yet, Wood says, 14th-century descriptions of the Black Death are vague. They're "usually too non-specific to be a reliable basis for diagnosis," he says.

IN CONCLUSION: So what really caused the Black Death?
Climate change, according to Wood and co-researcher
Sharon DeWitt. A global warming period ended about
1300, followed by a cooling period called the Little Ice
Age. During that time, rain and bad winters prevailed,
and crops began to die in 1315. As people started to starve,
malnutrition caused their immunities to drop. DeWitt said,
"The pattern we observed, of the Black Death targeting the
weak but also killing people who were otherwise healthy, is
consistent with an emerging disease striking a population
with no immunity." Lack of immunity to disease combined
with poor sanitation led to the fast spread of disease and the
huge number of deaths.

Mixed-Up Geography

11TH GRADE STUDENT: "Egypt really exists?
I thought it was just some place from
Jimmy Neutron [a cartoon series]."

ANOTHER STUDENT: "What do you mean *place*?
I thought Egypt was a religion."

ROBO JELLYFISH

Out of the millions of animals in the world, the jellyfish is among the weirdest. But their odd, alienlike anatomy and elegant movement through water are exactly why they became the basis for an aquatic drone.

Engineers at Virginia Tech built Cyro, a 170-pound robotic jellyfish, with $5 million in funding from the U.S. Naval Undersea Warfare Center and the Office of Naval Research. The navy hopes that this project will one day lead to autonomous underwater robots that are capable of underwater surveillance of both the environment and potential armed threats.

Cyro is equipped with a silicone cover over its metallic frame, which gives it its jellyfish camouflage. It also comes equipped with a rechargeable battery that gives it four hours of life, and a computer system that allows it to be programmed to perform individual missions. So if you bump into a jellyfish while swimming, check to make sure you got stung by a real jellyfish before asking someone to pee on you.

LIFE ON MARS?

For centuries, humans have looked up at our closest planetary neighbor and wondered if we would ever live there. Today, scientists are working on making this a reality. NASA has even announced a date for the first manned mission to the Red Planet: 2031.

The bad news: It may be closer to the year 3031 before a human can take a stroll around Mars wearing nothing but a pair of shorts and a T-shirt. As it stands right now, Mars's average temperature is −81°F, its atmosphere is extremely thin, and it contains almost no oxygen. To fix all three of these problems would be, by far, the largest and boldest undertaking in human history.

The good news: Mars possesses many of the basic elements necessary for life to develop, the most crucial being water. The planet also has a promising atmospheric makeup: 95.9 percent carbon dioxide, 1.9 percent nitrogen, and 0.15 percent oxygen. While that's far below the 20 percent oxygen in our atmosphere, it's encouraging because four billion years ago, Earth's atmosphere was nearly the same as Mars's is today. So, to make Mars earthlike—or *terraformed*—it needs heat, more water, a thicker atmosphere, and lots and lots of oxygen. But how do you do it in less than four billion years? One theory is on the next page...

SOLAR SAILING TO MARS

One of the ways that NASA may send humans to Mars is on a ship powered by "solar sails," giant mirrors that harness the sun's energy to propel the ship forward. This same principle could be used to heat Mars by reflecting sunlight to the surface. However, the mirrors would need to be about 150 miles wide to heat enough land to make it worthwhile. Mirrors that large couldn't be assembled on Earth, so the alternative would be to assemble them in orbit out of "space junk"—floating debris from previous space missions, jettisoned fuel tanks, and old satellites (now *that's* recycling). Once installed 300,000 miles above the Martian surface, they'd be trained on the frozen polar caps and begin melting the ice. This process would release CO_2 into the atmosphere, theoretically triggering the greenhouse effect: CO_2 absorbs the sun's radiation, and having more of it would warm the planet and thicken the atmosphere.

The moving gases from the melting ice caps would also generate planet-wide dust storms, increasing the temperature even more. Eventually, Mars would be warm

enough for liquid water to develop (but not freeze) at the poles. At this point, rockets filled with algae spores would be sent to this new ocean. The new algae would thrive in the water, causing photosynthesis, a by-product of which is oxygen. Humans would still need to wear air tanks for a few millennia, but the amount of oxygen would increase as the temperature slowly rose.

*Go to page 357 for another possible way
humans could colonize Mars.*

> "I seem to have been only like a boy playing on the seashore, and diverting myself in now and then finding a smoother pebble or a prettier shell than ordinary, whilst the great ocean of truth lay all undiscovered before me."
>
> **—Isaac Newton**

FRANKENFOODS

Farmers have been creating new kinds of plants for hundreds of years. In the past, they used an old-fashioned method: cross-pollination. That means they mixed the pollen of similar plants to create a hybrid, a new kind of plant. Scientists now breed plants by manipulating their genes to produce a genetically modified organism, or GMO. And they don't just cross one plant with another: they mix plant genes with animal genes. So scientists creating "transgenic" plants may be more like Dr. Frankenstein than Old MacDonald.

Here are a few examples of the "plants" they've already made, or have tried to create:

- Tomatoes that have genes from an Arctic flounder to make them resistant to frost.
- Corn crossed with genes from a bacterium to make it poisonous to insects.
- Apples with a gene taken from a moth to make the apple tree resistant to fire blight (a disease that destroys millions of dollars' worth of apples worldwide every year).
- Smart crops with a firefly gene that makes them glow when they need water.

In nature, transfer of genes happens only between closely related species. In a genetically engineered world—according to critics—nature as we know it might cease to exist.

HOW A MICROWAVE WORKS

- Like visible light, radio waves, and X-rays, microwaves are waves of electromagnetic energy. What makes the four waves different from each other? Each has a different wavelength and vibrates at a different frequency.

- Microwaves get their name because their wavelength is much shorter than electromagnetic waves that carry TV and radio signals.

- The microwaves in a microwave oven have a wavelength of about four inches, and they vibrate 2.5 billion times per second—about the same natural frequency as water molecules. That's what makes them so effective at heating food.

- A conventional oven heats the air in the oven, which then cooks the food. But microwaves cause water molecules in the food to vibrate at high speeds, creating heat. The heated water molecules are what cook the food.

- Glass, ceramic, and plastic plates contain virtually no water molecules, which is why they don't heat up in the microwave.

- When the microwave oven is turned on, electricity passes through the magnetron, the tube that produces microwaves. The microwaves are then channeled down a metal tube (waveguide) and through a slow rotating metal fan (stirrer), which scatters them into the part of the oven where the food is placed.

- The walls of the oven are made of metal, which reflects microwaves the same way that a mirror reflects visible light. So when the microwaves hit the stirrer and are scattered into the food chamber, they bounce off the metal walls and penetrate the food from every direction. A rotating turntable helps food cook more evenly.

- Do microwave ovens cook food from the inside out? Some people think so, but the answer seems to be no. Microwaves cook food from the outside in, like conventional ovens. But the microwave energy only penetrates about an inch into the food. The heat that's created by the water molecules then penetrates deeper into the food, cooking it all the way through. This secondary cooking process is known as conduction.

- The metal holes in the glass door of the microwave oven are large enough to let out visible light (which has a small wavelength), but too small to allow the microwaves (which have a larger wavelength) to escape—so you can see what's cooking without getting cooked yourself.

The Rock Painting Hoax

BACKGROUND: A new piece of ancient artwork turned up in the British Museum in 2005. The artifact was a rock bearing painted images of animals, a man, and an unusual tool. The sign beneath it read: "Early man venturing towards the out-of-town hunting grounds."

EXPOSED! The "tool" in the picture was a shopping cart; the "artifact" had been secretly placed there by British hoax artist Banksymus Maximus, also known as "Banksy." He designed it to look like the authentic ancient pieces in the museum—and it stayed up for three days before "experts" at the museum noticed it. (The sign on the piece also dated it to "the Post-Catatonic era.") The museum took the hoax in good humor, and even returned the piece to the artist. It quickly went up at Banksy's latest show at another museum, with the label "On loan from the British Museum."

Love Potion #9

The word *aphrodisiac* comes from Aphrodite, the Greek goddess of love and beauty. Eventually, according to the *Dictionary of Word and Phrase Origins*, her name was used "to describe any drug or other substances used to heighten one's amatory desires." Here are some aphrodisiacs used throughout history.

- In the Middle Ages people believed that "eating an apple soaked in your lover's armpit is a sure means of seduction." Others drank the urine of powerful animals to increase sexual powers.

- A 15th-century Middle Eastern book entitled *The Perfumed Garden for the Soul's Delectation* suggested that lovers eat a sparrow's tongue, and chase it down with a cocktail made of honey, 20 almonds, and parts of a pine tree.

- People once thought that eating any plant that looks phallic would increase male virility—carrots, asparagus, and mandrake root were especially popular. Bulbs and tubers—e.g., onions—which people thought resembled testicles, were also believed to increase sexual potency. And peaches, tomatoes, mangos, or other soft, moist fruits were considered aphrodisiacs for women.

- In *Consuming Passions*, Peter Farb and George Armelagos write that during the 1500s and 1600s, "Europe was suddenly flooded with exotic plants whose very strangeness suggested the existence of secret powers." For example:

  Tomatoes brought back from South America were at first thought to be the forbidden fruit of Eden, and were known as "love apples." And when potatoes first arrived in Europe—the sweet potato probably brought back by Columbus and the white potato somewhat later—they were immediately celebrated as potent sexual stimulants...A work dated 1850 tells the English reader that the white potato will "incite to Venus."

- In the 20th century, everything from green M&M's to products like Cleopatra Oil and Indian Love Powder have been passed off as aphrodisiacs. Even in 1989, a British mail-order firm called Comet Scientific was offering an aerosol spray that it claimed made men "irresistible to women."

> When you see something you like, your pupils dilate.

THE DANCE OF THE DUNG BEETLE, PART 1

What's cuter than a dung beetle wearing tiny boots? Lots of things, actually, but according to biologist Eric Warrant at Sweden's Lund University, "They're the cutest animals you can imagine"; he also admits that "most people find them a bit revolting."

Cute or not, these are no ordinary bugs. In a 2012 study, the Swedish research team discovered something new about dung beetles: they dance on top of their balls of dung to cool off from the African heat, which can reach 140°F. And yes, they discovered this by placing tiny silicone boots on a group of test beetles, which kept their feet cooler. The booted beetles spent much less time dancing on the ball than the barefooted beetles.

Why is this significant? Just as Jane Goodall's chimpanzee research taught us that humans aren't the only tool-making mammals, "Dung beetles are the first example of an insect using a mobile, thermal refuge to move across hot soil," said lead researcher Jochen Smolka. "Insects, once thought to be at the mercy of environmental temperatures, use sophisticated behavioral strategies to regulate their body temperatures."

So now we ask: Is there anything cuter than a dung beetle wearing tiny boots? Go to the next page to find out.

THE DANCE OF THE DUNG BEETLE, PART 2

It's long been known the male dung beetle rolls mammal feces into a ball several times larger than itself and then rolls it backward to its hole, where it will feed the entire beetle family. It's also been known that beetles, like all insects, use the sun to navigate, but they've also been observed rolling the dung balls at night, even when there's no moon. How do they know where to go? That's where the tiny hats come in. In 2012, researchers led by Lund University's Marie Dacke set up a testing area where the beetles couldn't see the horizon. Then they put tiny hats on some of the beetles that blocked out the stars. Those beetles didn't know where to go. The researchers discovered that they were navigating by the band of stars we call the Milky Way galaxy, which is much brighter and more prominent in the southern African sky. Not bad for a dung beetle.

THE MOON CAME FROM THE PACIFIC

Over the centuries, there have been several theories explaining how the Moon came to be. Three prevailed for more than 100 years, until humans actually traveled to the Moon in the 1960s.

The Co-Accretion Theory (late 1800s). The exact origin of this hypothesis is unknown, but it was a popular one. It proposed that the Moon was formed during the *accretion* process that formed Earth. For this theory to be true, however, the Moon would have to have the same chemical composition as Earth— and lunar samples taken during the Apollo missions showed that this is simply not true.

The Fission Theory (1878). Mathematician and astronomer George Howard Darwin, son of Charles Darwin, proposed that centrifugal force caused by the spinning of the very early, still molten Earth caused a large piece of it to break off and fly into space, where it was caught in orbit and became the Moon. In 1882 geologist Osmond Fisher added that the Pacific Ocean basin was the scar left behind by this event. Again, lunar composition makes this theory very unlikely.

The Capture Theory (1909). This theory was proposed by American astronomer Thomas Jefferson Jackson See. It says that the Moon was formed far away from Earth and later passed close enough to be captured in its gravitational field. That would explain why the composition of the Moon is different from Earth, but this theory is largely discounted due to the extreme unlikeliness that a passing object would be moving at the correct speed and direction to be pulled into a perfect orbit.

Go to page 266 for more about the Moon's origin.

Scientists at Birkbeck College in England discovered that, like humans, dogs can "catch" yawns from people. A 29-dog study found that after they made eye contact with a yawning person, 21 of the dogs yawned as well.

SNAPSHOT OF SCIENCE
WACKY GENIUS
Princeton, New Jersey, March 14, 1951

Many people know that Albert Einstein's brain was preserved after his death. Fortunately, his tongue was preserved, too—on film—and made him everyone's favorite genius. The photo was taken on the campus of Princeton University as the famed physicist was celebrating his 72nd birthday. Asked to smile for the umpteenth time, he stuck out his tongue instead and photographer Arthur Sasse captured it in full extension.

Einstein was celebrated for the wild tangle of his long hair and his expressive face, or as one put it, "a cartoonist's dream come true." In fact, the scientist became the model of the mad scientist (or absentminded professor) as depicted in the film *Back to the Future.* But that pointed tongue, at full extension, became the iconic poster of countless college dorm rooms. We can't be sure, but perhaps it was inspiration for the rock group Kiss and bass guitarist Gene Simmons, whose "ten-foot tongue seen around the world" is the group's enduring symbol.

STRANGE SCIENCE

FIVE FREAKY FACTS ABOUT...
TESLA

- Nikola Tesla was born in 1856 during a lightning storm. While growing up in Croatia, he complained of "blinding flashes of light" followed by visions that he later said inspired his greatest inventions.

- Tesla was afraid of round things, especially pearls. He detested pearls so much that he wouldn't speak to a woman who was wearing them.

- Tesla could fluently speak eight languages: Serbo-Croatian, Czech, English, French, German, Hungarian, Italian, and Latin.

- The more you use your brain, the lighter your eye color becomes...according to Tesla. He said that a life of science had lightened his eyes.

- Tesla had an odd fascination with the number three that lasted until the end of his life...in 1943, three days before what would have been his 87th birthday (divisible by 3) in room 3327 (also divisible by 3) on the 33rd floor of his hotel.

More "Science" Museums

TATTOO ART MUSEUM

Location: San Francisco, California

Background: This museum is housed in the working tattoo parlor of owner Lyle Tuttle. Learn about the history of the tattoo in different societies, such as Japanese, Samoan, and Native American. But be forewarned: the collection of skin-engraving paraphernalia might make you think twice about that cute butterfly you were thinking about putting on your ankle.

FUTURE BIRTHPLACE OF CAPTAIN KIRK

Location: Riverside, Iowa

Background: Boldly going beyond a single building, the entire town is a *Star Trek* museum. According to the TV series, Captain Kirk will be born here on March 22, 2233. The town wanted to erect a statue of William Shatner as Kirk, but Paramount Pictures, which owns *Star Trek*, wanted a $40,000 licensing fee. So instead, docked in the town square is the USS *Riverside*, which bears a striking (but not copyright-infringing) resemblance to the USS *Enterprise*. For $3 you can buy Kirk Dirt, a vial of soil dug from his future birth site.

THE MAGIC MAN

John Dee (1527–1608) was a mathematician, alchemist, academic, and astronomer in Britain. From an early age Dee displayed two conflicting sides of his character: as a scientist, his feet were firmly planted on the ground; as a mystic, he *really* had his head in the clouds. Dee's academic reputation grew quickly and he was offered (but declined) prestigious professorships in Paris and at Oxford.

Then, in 1553, Queen Mary (Bloody Mary) ascended the British throne. John Dee found himself on her hit list. According to the thinking of the time, supernatural spells and spirits could be conjured through scientific and mathematical experimentation. Authorities targeted scientists and mathematicians. Dee was arrested in May 1555 and held for three months before being released.

In 1558 Bloody Mary died and was replaced as queen by Elizabeth I. Dee was quickly back in favor. The new queen asked Dee to use his astrological skills to divine the most auspicious day for her official coronation, and made him her court astrologer. Why the dramatic turn? Many historians believe Dee acted as a spy for Elizabeth during her sister Mary's reign, keeping her informed of which academic colleagues were supportive of Mary. The theory is difficult to prove or disprove.

Dee continued to look for answers to the mysteries of the universe. By 1566 he had collected innumerable scientific instruments and the largest library of books in Britain, attracting visitors from all over.

HOW TO WIN A NOBEL PRIZE

You can't nominate yourself. If you do, you'll be automatically disqualified. No exceptions!

You must be alive. Nominating dead people is not allowed. If you die, you're out of the running. Even if no one else was nominated.

There are no runners-up. If you come in second to someone who drops dead before he or she picks up the medal, you still lose.

You can't win by default. If you come in second to someone who refuses the medal, do you win, or at least get the prize money? No and no.

No organizations allowed. With the exception of the Nobel Peace Prize, no single prize can be awarded to more than three people.

You don't get a laurel. The term "Nobel laureate" is just an expression. If you win, you get a gold medal, a diploma with your name on it, and a cash prize (about $1 million). If you want to go around wearing a laurel-leaf crown like Julius Caesar, you'll have to make it yourself.

HAT TRICK

On page 92, we told you about black hats,
a type of computer hacker. Here's just one of many
things they have the potential to do.

At the 2010 Black Hat Security Conference, security
researcher Barnaby Jack wowed his audience by hacking
into two different ATMs right from the stage. He used a
remote connection for one and a USB port on the other,
and made them both spit out money like a Las Vegas
slot machine. How'd he do it? He wouldn't go into detail
(otherwise we'd all become Black Hats looking for a
jackpot), but he made it clear that it wasn't just ATMs that
were vulnerable. Every piece of equipment that uses a
standard computer, like the kind inside an ATM, can be
easily hacked: cars, medical devices, televisions, you name
it. Jack also pointed out that once he hacked a bank ATM,
the machine's data gave him access to anyone who'd ever
used it. He found that the ATMs at convenience stores were
the easiest to hack.

WHAT RACE(S) ARE YOU?

Recent breakthroughs in the science of genetics have had a huge effect on the world, with applications in medicine, agriculture, law enforcement, and more. *Genetic markers* found through DNA tests can also reveal familial and ethnic lineages. And the science behind it, while immensely complex in its details, is pretty simple.

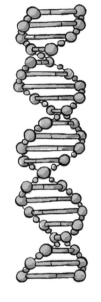

Human DNA is alike in every person—but it's not *exactly* alike. Individuals can acquire *mutations* along the way. Some genetic mutations cause disease, some affect eye or hair color, some do nothing at all. The ones that are used for racial testing are, primarily, ones that have no known effects. Say a guy named Bob acquired such a mutation. Bob had 10 kids—and he passed that mutation down to them. They each had 10 kids—and they all got the mutation, too. This kept happening over many generations, and today there are tens of thousands of people with that specific "Bob" mutation. And they're the only ones on earth that have it. Well, that's exactly what happened throughout human history.

DNA and fossil evidence suggests that modern *Homo sapiens* first appeared in northeastern Africa roughly 200,000 years ago. Their descendants began migrating out of Africa about 60,000 years ago, spreading in different directions. Travel then was difficult, so those different groups of travelers didn't interact for a very long time. The people who would go on to become the Native Americans, for example, wouldn't interact with the people who went on to become the Europeans for many thousands of years. That long genetic separation resulted in entire groups acquiring DNA mutations that were unique to them. The people who became the Native Americans acquired mutations different from those the Europeans acquired. When means of travel progressed and long-separated groups of peoples *did* start interacting, and having children together, those mutations started being shared. And now— we can find them.

•••••••••●•••••••••

THAT'S NO BROOMSTICK

Modeled on Harry Potter's flying broomstick, Mattel decided to add a touch of realism to the Nimbus 2000 by adding a battery-powered motor that made the broomstick "simulate movement"...by vibrating. To recap: That's a foot-long toy meant to be stuck between the legs that *vibrates*. The toy was quickly discontinued.

WEIRD DYES

Three thousand years ago, fishermen in the eastern Mediterranean found a sea snail called a *murex* that had an unusual property: if you squeezed its sluglike body, it oozed a deep, brilliant purple substance that made a beautiful dye. The color became so prized that the area became known as Phoenicia, the "Land of Purple." The dye was named "Tyrian purple" after Tyre, the capital of Phoenicia.

It took nearly six million snails to make a single pound of dye, and a single *ounce* of dye cost a pound of gold. Only the wealthy could afford purple clothes, and purple became the royal color of the empires of Egypt, Persia, and Rome.

By the fifth century, the murex snail population was on the verge of extinction. Kings and popes needed alternatives to Tyrian purple. In 1464 the Catholic Church introduced "cardinal's purple," a maroon dye made from an insect called a *kermes* (which gave its name to the new dye's color, *carmine*). A hundred years later, the Spanish brought another insect-derived dye—cochineal—back from Mexico and Peru. The Aztec king Montezuma had worn robes dyed in this brilliant red. Cochineal is still used today in food coloring, cosmetics, and more.

JOURNEY INTO SPACE

Journey into Space was a BBC radio show that aired from 1953 to 1958. Unlike most science fiction programs that were complete flights of fancy, this show was grounded in the real physics of spaceflight. In one episode a group of reporters is given a tour of a launchpad on the Moon; the description of the spacecraft is so true to life that the modern listener may forget that the show predated the *Apollo* Moon landing by 15 years. The realism helped make it one of most listened-to radio series in the history of the BBC, and the last one to attract a larger audience than the television shows that were on at the same time. The episodes are now available on CD.

THINGS TO LISTEN FOR: Lemmy, the clueless Cockney member of the crew. He has presumably spent years training for the first mission to the Moon in episode 1, yet he is surprised to find out that he is weightless in space. Why? In the early 1950s, most listeners had no understanding of spaceflight; having someone explain it to Lemmy was the show's way of explaining it to the audience.

A VISIT TO THE WITCH DOCTOR

In the Zulu, Swazi, Xhosa, and Ndebele tribal traditions of South Africa, *sangoma* are healers who call upon dead ancestors to help diagnose and cure patients. Most sangoma cures are made with ingredients such as plants. But cures like the ones below might be a bit harder to swallow.

PROBLEM: Backache
CURE: Eat crocodile fat.

PROBLEM: You want to win a soccer game.
CURE: Eat lion fat.

PROBLEM: Your crops won't grow.
CURE: Bury a human skull in the field.

PROBLEM: You want to be elected president.
CURE: Eat a slice of brain.

PROBLEM: Stroke
CURE: Eat lizard flesh.

PROBLEM: You feel weak.
CURE: Eat ground-up finger bones.

PROBLEM: You're sleepy and low on energy.
CURE: Drink human blood.

PROBLEM: Your store needs more customers.
CURE: Bury a human hand under the entrance.

Who Are the Biohackers?

Depending on who you ask, biohackers are either courageous "citizen scientists" who improve the human condition through genetics and technology...or they're reckless wannabes who lack any respect for the scientific method. Whoever they are, biohackers are into some pretty strange stuff, such as...

UPGRADED FOCUS BRAIN TRAINER: Developed by biohacker Dave Asprey, this "near infrared, hemoencephalogography device feedback system" is a headband that measures the flow of blood going into your brain. When your readings are low, you can do brain exercises to get the blood flowing again.

CRISPR-CAS9: Short for "Clustered Regularly Interspaced Short Palindromic Repeats," Crispr-Cas9 was developed by Sebastian Cocioba, a 25-year-old college dropout who set up a lab in his parents' New York apartment. The Cas9 enzyme "acts as a pair of 'molecular scissors' that can cut the two strands of DNA at a specific location in the genome so that bits of DNA can then be added or removed." The implications of this tech could lead to everything from designer babies to genetic weapons of mass destruction.

DOING SCIENCE IN THE DARK

In 2015 Gabriel Licina and Jeffrey Tibbetts—two biohackers from a California collective called Science for the Masses (SM)—performed the first human trial of a "chlorophyll analog" called Chlorin e6 (Ce6) that they claim gives the user night vision. Tibbetts dropped some in Licina's eyes, and then he and four control subjects went to a dark forest to see what they could see. According to their (non-peer-reviewed) paper (basically a blog post), SM concluded, "The Ce6 subject consistently recognized symbols that did not seem to be visible to the controls."

After several media outlets breathlessly reported "Biohackers create night vision superpowers!" several actual scientists called out SM for its unorthodox methods—such as not using double-blind control subjects. "Unfortunately, due to the weak design of this experiment," wrote Peter Rothman in *Humanity+* magazine, "we really cannot conclude anything conclusively about the true effect of Ce6 on night vision." Rothman pointed out that Ce6 wasn't even invented by SM; it was patented in 2006 by Dr. Ilyas Washington, who wrote, "This mechanism is shown to enhance vision in a mouse model and perhaps could also do so in humans."

SM then took it upon themselves to formulate their serum. After a "totally disgusted" commenter complained that "research on human subjects requires FDA and Ethic Committees review," Licina rebutted that he doesn't need permission if he himself is the test subject, adding, "We're terribly sorry that you didn't find a post doc degree attached to our names which is, I am assuming, the magic pass that allows people to play around in a lab."

DOUBLE EXTINCTION

Paleontologist Othniel Charles Marsh (1831–1899) was the first to describe and name the *Apatosaurus* dinosaur, based on the discovery of a few bones in Colorado. Then one of his teams found an almost complete (and what seemed like a distinctively new) skeleton in Wyoming. Marsh plunked a random head on his headless skeleton and named it *Brontosaurus*. This conveniently got him credit for two discoveries. Eventually, though, other paleontologists figured out that the two skeletons were from the same dinosaur and that Marsh's *Brontosaurus* had just been an adult *Apatosaurus*. The name *Brontosaurus* was formally discarded in 1974, thus making the *Brontosaurus* extinct...again.

Oh, Baby!

Sausage isn't the healthiest of foods. Pork and beef sausages in particular are often made from the fattiest parts of the animal, and they're loaded with saturated fat. What if food scientists could create a sausage that was a little bit better for you, in spite of all that fat? They have. Researchers in Spain have concocted sausage laden with digestion-friendly probiotics—probiotics harvested from the poop of babies.

Generally, a baby's digestive tract is clean and healthy, not yet burdened by years of use and poor dietary habits. Scientists at the Institute of Food and Agricultural Research found that microbes in infant poop (who knows what made them think to look *there*) contained large amounts of two probiotics that are almost nonexistent in adult poop.

Sausage is cured with bacteria naturally present in the animal flesh being used, so the scientists took the baby poop bacteria (harvested from 43 babies no older than six months old) and used that as the fermenting agent instead. Result: baby-poop sausage. Team member Anna Jofré claims that the sausage is good for the lactose intolerant—dairy products are a primary source of probiotics, particularly yogurt. "Probiotic fermented sausages will give an opportunity to consumers who don't take dairy products the possibility to include probiotic foods to their diet." Jofré also assured a reporter that the sausages "tasted very good."

ANALYTICAL ADA

When Annabella Milbanke married the poet Lord Byron in 1815, he was already famous as the creator of the brooding, defiant romantic hero. A year after they were married, Lady Byron went home to her parents, taking baby daughter Ada with her. Taking no chances that her daughter might grow up to be a poet, Lady Byron hired a series of tutors to educate little Ada in mathematics and science, as well as reading and writing. By the time she was 13, Ada knew more about math than her tutors did.

ANALYTICAL ENGINE

At a dinner party one night in 1833, Ada—by then married to the Earl of Lovelace—heard inventor Charles Babbage talk about his calculating machine: the analytical engine. Though neither she nor Babbage would ever see it finished in their lifetimes, they both understood how it could work.

Ada translated an Italian article about it, and added a footnote of her own that discussed the difference between a simple calculating machine and the analytical engine—a difference like that between pocket calculators and computers today. Babbage suggested she add more of her own ideas, which turned out to be three times as long as the original article and was published in

1843 in a serious science journal. Her powerful imagination allowed her to make leaps beyond the available information. Aware that Babbage had based his designs on a weaving loom, she wrote in a letter, "The Analytical Engine weaves algebraic patterns just as the Jacquard loom weaves flowers and leaves."

CREATING A COMPUTER IN HER MIND

Lady Lovelace described how the engine could produce important number sequences, deal with symbolic sequences (like algebra), have a memory, and how subroutines could be built in for special tasks. She predicted it could compose music and produce graphics. She even considered artificial intelligence, and explained why A.I. wouldn't work. In fact, her ideas were so good that a lot of people consider her the first computer programmer. (A 1979 U.S. Department of Defense software language was named in honor of her.)

Mixed-Up Earth Science

"Where do clouds go at night?"

—Asked by a high school student

RANDOM ORIGIN

NATIONAL GEOGRAPHIC MAGAZINE

In January 1888, thirty-three men (including world-renowned explorers, military officers, academics, bankers, and mapmakers) met at the Cosmos Club in Washington, D.C., to organize a group whose mission was to "increase geographical knowledge." Two weeks later the National Geographic Society was officially established, and the first issue of *National Geographic Magazine* was published in September 1888. It was a dry, academic journal, but still attracted readers thanks to photographs from exotic places as well as maps and archaeology reports.

It didn't become the magazine it is today until Alexander Graham Bell was named president of the society in 1897. Among Bell's innovations: He had the magazine printed on thick paper so it felt more like a book, devised the yellow-trimmed photographic cover, and solicited rollicking firsthand accounts from explorers like Robert Peary and Ernest Shackleton. He also realized that the magazine's strength was showcasing photos. By 1908 photos took up half of the magazine, and even more than that after *National Geographic* ran color images in the early 1930s. By 1950 it was one of the top 10 most-read magazines in the world. It now reaches more than 50 million readers every month.

DREAM DISCOVERY: LEAD SHOT

British plumber William Watts came up with the process for making lead shot used in shotguns. This process was revealed to him in a dream. At the time, making the shot was costly and unpredictable—the lead was rolled into sheets by hand, then chopped into bits. Watt had the same dream each night for a week: He was walking along in a heavy rainstorm—but instead of rain, he was being showered with tiny pellets of lead, which he could see rolling around his feet. The dream haunted him; did it mean that molten lead falling through the air would harden into round pellets? He decided to experiment. He melted a few pounds of lead and tossed it out of the tower of a church that had a water-filled moat at its base. When he removed the lead from the water, he found that it *had* hardened into tiny globules. To this day, lead shot is made by being dropped from a height into water.

WEIRD ENERGY:

SOLAR WIND

As long as the Sun shines, solar power may be an infinitely renewable resource. But the Sun provides another source of potential usable power in the form of solar wind. Solar wind is a high-powered stream of charged particles constantly shooting out of the Sun. Brooks Harrop and Dirk Schulze-Makuch of Washington State University are the leading researchers on the idea, and they believe that a Sun-orbiting satellite could be used to capture those beams of energy. Using solar-powered batteries to run an electric charge through a copper wire, the satellite would generate a magnetic field that would in turn attract solar wind particles. The energy could then be zapped to a receiver on Earth via an infrared laser. While all this sounds like science fiction, the principles are scientifically sound. The main problem Harrop and Schulze-Makuch are trying to solve is how to aim and shoot a laser beam from the Sun to the Earth—a distance of nearly 100 million miles—without losing much energy. Harrop and Schulze-Makuch think their technology could at least be used to beam solar wind energy to other satellites and spacecraft. How much energy could solar wind ultimately provide? One hundred billion times the planet's current power needs.

The Great Moon Hoax

PERPETRATOR: The *New York Sun* newspaper

STORY: The paper printed its first issue in 1833, and by 1835, it was looking for a circulation boost. So to drum up interest, editors announced the upcoming publication of six articles covering renowned British astronomer Sir John Herschel's fantastic new "discoveries" of life on the Moon: forests and seas, cranes and pelicans, herds of bison and goats, flocks of blue unicorns, sapphire temples with 70-foot pillars—even a race of batlike humanoid creatures. According to the *Sun*, the articles would be reprinted from the *Edinburgh Journal* of *Science*.

The day the first article appeared, *Sun* sales were 15,000; by the sixth day, they were over 19,000, the highest of any New York paper at the time. Other newspapers, racing to catch up, claimed to have the "original" *Edinburgh Journal* articles too, but they actually just reprinted the *Sun*'s stories.

EXPOSED: There were no *Edinburgh Journal* articles. In fact, that journal had gone out of business several years earlier. And Herschel, perhaps the most eminent astronomer of his time, was totally ignorant of the hoax (and then amused by it until he got sick of answering questions about Moon men). The articles were reportedly written by *Sun* reporter Richard Adams Locke. The *Sun* never formally admitted the deception, but it did publish a column speculating that a hoax was "possible." Regardless, the paper got what it wanted and circulation remained high.

STRANGE SCIENCE

FIVE FREAKY FACTS ABOUT...
GLOBAL RESEARCH

- Italian doctor Gabriele Falloppio conducted the first clinical trial of condoms in 1546. He made them himself out of linen, and they were meant to prevent syphilis, not pregnancy.

- In the 1920s, Swedish zoologist Sten Bergman discovered a subspecies of bear in northeast Russia that was much larger and darker than common brown bears. But Bergman never actually spotted one; he saw only a hide and tracks. No one has ever seen what is known as Bergman's bear.

- A study in Spain determined that rats can tell the difference between the Dutch language and Japanese.

- Japanese researchers found that tiny snails eaten by birds can pass through their gut and come out alive. According to *National Geographic Daily News*, "One snail even gave birth shortly after emerging— apparently unfazed by its incredible journey."

- In 2015, an astrophysicist at the Parkes radio telescope in Australia discovered the source of strange signals that had mystified her peers for 17 years. It was a microwave in the facility's kitchen.

HOW AN X-RAY MACHINE WORKS

An X-ray machine is basically a camera, but it creates its own "light" in the form of X-rays. An electric current is sent to the vacuum tube that houses a *cathode* at one end and an *anode* at the other. The cathode is a filament, like in lightbulbs. When the current passes through it, it heats up and emits electrons—negatively charged atomic particles—into the vacuum tube. The anode is positively charged, so it acts as a magnet and pulls those negatively charged electrons toward it. Embedded in the anode is a metal disk, usually made of tungsten. When one of those incoming electrons collides with a tungsten atom, that atom loses one of its own electrons—so another electron in the atom jumps in to fill it.

The electron that jumps in comes from farther away from the atom's nucleus—and has much more energy than the one that got knocked out. When it fills in the empty spot, it has to release its surplus energy—which it releases as a *photon*. This happens billions of times in one X-ray procedure.

A lead container surrounds the vacuum tube to absorb X-rays. A small opening in the container allows the escape of photons, which first pass through a series of filters that stop all but a pure beam of X-ray photons.

Now we take the picture. On one side of whatever is being X-rayed—say your hand—is the window through which the beam is emitted; on the other side of your hand is the film. As the X-ray photons meet the atoms in your hand, they will either pass through or be absorbed by them. The softer tissues in your body—your skin, organs, and muscles—are made up of relatively small atoms. The X-ray photons pass right through them and leave their mark as the lighter areas on the film, just like light hitting a camera's film. The atoms in your bones—primarily calcium—are much larger, and they absorb the X-rays and stop them from reaching the film. Result: The film has captured an image of your bones.

Green City:
COPENHAGEN

POPULATION: 1.2 million

HOW GREEN IS IT? Copenhagen, Denmark, has been addressing environmental issues for decades. The result is that the water in its harbors and canals is so clean that people actually swim in them. There are also more than 186 miles of bike paths in the metro area, and residents and tourists can borrow bikes for free. Some major streets even have a "green wave" system so bike riders can speed through intersections without stopping—they hit timed green lights the entire way. Now more than half of Copenhageners bike to work or school.

The city is filled with parks, and plans are in the works to guarantee that at least 90 percent of Copenhagen's population will be within walking distance of a park or beach. About 20 percent of the city's electric power comes from wind turbines, hydroelectric power, and biomass (energy from organic matter like wood, straw, and organic waste), but the goal is to stop using coal altogether. The city is encouraging residents to buy electric- and hydrogen-powered cars and is investing more than $900 billion so that, by 2025, Copenhagen will be carbon-neutral.

IT'S REIGNING DINOSAURS

Test your knowledge of these prehistoric creatures.

1. Some 2,000 years ago, dinosaur fossils were discovered in Wucheng, Sichuan, China. What did people think they'd found?

2. Most scientists believe Dr. Gauthier's theory that these modern-day animals are the descendants of dinosaurs.

3. Paleontologist Jack Horner discovered the first evidence that some dinosaurs do what?

4. In 2016 scientists confirmed that a skeleton nicknamed "Wade" is in fact a new species of titanosaur. In what country was it found?

5. How many books did Michael Crichton write in his dinosaur thriller series that was made into four movies?

6. Able to look into a six-story window, what dinosaur was likely the tallest creature to walk the earth?

7. "Sue," sold for more than $8 million to the Field Museum of National History in Chicago, is famous because it is the most complete skeleton found of what dinosaur?

8. Who wrote, "If we measured success by longevity, then dinosaurs must rank as the number one success story in the history of land life"?

9. What dinosaur, related to Tyrannosaurus rex, was found in and named for a Canadian province?

10. Megalosaurus, the first dinosaur fossil to get a scientific identification, was discovered in 1824 in what country?

11. What dinosaur name means "fast thief"?

ANSWERS: 1. Dragon bones **2.** Birds **3.** Care for their young **4.** Australia **5.** Two: *Jurassic Park* and *The Lost World* **6.** *Sauroposidon* **7.** *Tyrannosaurus rex* **8.** Robert T. Bakker **9.** *Albertosaurus* **10.** England **11.** *Velociraptor*

Titanosaurus Femur Bone

REAL-LIFE TIME MACHINE

A 2015 news report told of Ronald Mallett, a professor at the University of Connecticut who has dreamed of building a time machine since he was 10. That's when his father died of a heart attack. Mallett figured that if he could travel back in time, he could warn his father to stop smoking.

Mallett's time-machine design uses light energy (four laser beams) to warp space and time. It takes a PhD to understand the details, but according to Mallett, the beams could swirl space and time like "a spoon stirring milk into coffee." Mallett says he'll need about 10 years in the lab to make his time-travel dream real.

Almost all flu viruses first infect chickens, then pigs, and then spread to humans, mutating along the way. But the chicken flu of 1997 made medical news because it jumped directly from birds to humans, bypassing pigs.

DUMB PREDICTIONS

"The wireless music box has no imaginable commercial value. Who would pay for a message sent to nobody in particular?"

—ASSOCIATES OF NBC PRESIDENT DAVID SARNOFF
(responding to his recommendation, in the 1920s, that they invest in radio)

The Mystery of the Stradivarius

Science has finally solved a mystery that has baffled luthiers (stringed instrument makers) for decades: why does a 300-year-old Stradivarius violin sound so much more "brilliant"—a musical term describing clarity of tone—than modern instruments?

Hwan-Ching Tai, a professor of chemistry at National Taiwan University, studied several instruments made by Antonio Stradivari in Cremona, Italy, and discovered that the wood was presoaked in a chemical solution containing aluminum, calcium, copper, and other minerals (most likely to ward off fungus). That practice is long gone, and today's violins have no such chemical additives... which is what Tai believes gives the old instruments their distinctively brilliant tones. Another factor: over the centuries, a compound in the wood called hemicellulose has mostly evaporated. That compound absorbs water, so less of it means less moisture, and a dryer wood gives a crisper sound.

So the good news is that science has finally solved the mystery; the bad news is that the only way today's luthiers could re-create the sound of a Stradivarius is to go back in time and work alongside him.

GREENHOUSE HELMET

Some of the world's biggest and best cities also have some of the world's biggest and best parks, perfect for jogging or bicycling. The problem is that those big cities are often horribly smog-choked, particularly on warm, sunny days; times that are preferred for outdoor exercise are rendered toxic by air pollution. The solution: the Greenhouse Helmet.

More than just a plastic bubble protecting you and your lungs from the bad air outside, it's a complete mini ecosystem inside. It consists of a large plastic dome that seals firmly around the head. Inside are tiny shelves outfitted with at least one plant, which, you might remember from junior-high science class, takes in the carbon dioxide humans expel, and in turn expels oxygen. This creates an environment of a constant exchange of clean air. And so you don't have to come across as a complete weirdo, the Greenhouse Helmet comes with speakers and a microphone for communicating with people who are not inside the Greenhouse Helmet. (Note: If you make one of these for yourself, be sure to poke some holes in it so you can breathe!) Though there was a U.S. patent application filed in the 1980s, it lapsed less than 10 years later.

MANHATTANHENGE

Most of the time, the tall buildings in Manhattan block the sunset. But twice a year, above 14th Street, the sun aligns with the streets' east–west grid pattern and sets perfectly between the buildings. It lasts only about 15 minutes, but it's so striking that people stop on the streets to watch. As solar rays light up the towering buildings, a glowing orange light filters along the streets. The reflection off the buildings also scatters the sunshine, sending bright light along the north–south avenues. Because the phenomenon resembles sunsets seen at England's mysterious Stonehenge ruins, astrophysicist Neil deGrasse Tyson calls it "Manhattanhenge."

Stonehenge was built by the ancient Celtic Druids to mark the exact moment of the spring and fall equinoxes. But Manhattan's street grid was established in 1811 for efficiency, not science, so it's slightly off center—it's turned 28.9 degrees from true east and west. As a result, the city's "equinoxes" occur on different days each year. Usually, the dates are in late May and mid-July. But if you miss the exact dates, not to worry. The days before and after Manhattanhenge also create a celestial glow—it's not quite as magnificent, but still pretty good.

According to Tyson, the best way to see Manhattanhenge is to position yourself as far east in Manhattan as possible, but ensure that when you look west across the avenues you can still see New Jersey. Clear cross streets include 14th, 23rd, 34th, 42nd, 57th, and several streets adjacent to them. The Empire State Building and the Chrysler Building render 34th and 42nd Streets especially striking vistas.

"If I had thought about it, I wouldn't have done the experiment. The literature was full of examples that said you can't do this."
—**Spencer Silver**
(on the adhesive that led to 3-M Post-Its)

BRUSH TALKS

Shen Kuo, raised in China in the 11th century, was successful in every role he took on: engineer, financial expert, military leader, manager, inventor, mapmaker, diplomat, and imperial academician. In his spare time, he studied math and astronomy at the Imperial Palace library.

At age 58, Shen retired to his garden estate. He called it Dream Brook because it matched a beautiful place he'd repeatedly seen in his dreams. Keeping to himself, Shen set to work writing a book of observations and thoughts that contained more than 600 essays. "I had only my writing brush and ink slab to converse with," said Shen. He titled the book *Brush Talks from Dream Brook.*

About a third of *Brush Talks* was science — biology, geography, physics, medicine, geology, archaeology, chemistry, astronomy, meteorology, and mechanics. But there were stories, too. One entry was titled "Purple Aunty Goes Down to the Mortal World." It described the custom of inviting the toilet goddess, Purple Aunty, into one's home on the eve of the Lantern Festival (the final day of Chinese New Year celebrations). Purple Aunty had the power to turn people into writers and poets.

Shen's book also included some remarkable firsts. His discussion of compasses noted that magnetized needles pointed to a spot that wasn't the exact North Pole, and he described a printing method using movable type—both 400 years before Europeans were credited with the discoveries.

STRANGE SCIENCE

The Four Ethnic Groups

Through relatively inexpensive DNA tests, people can learn valuable information about where their ancestors came from. One kind of test shows "genetic percentages" of ethnic heritage. A result of such a test might say that someone has 40% European, 40% African, 18% East Asian, and 2% Native American heritage. Or it might say 90% Native American and 10% East Asian.

The four very broad ethnic groupings determined by the test:

1. **African:** Peoples of sub-Saharan Africa.

2. **European:** This group is much broader than it appears, and includes peoples from Europe, North Africa, the Middle East, India, Pakistan, and Sri Lanka.

3. **East Asian:** Japanese, Chinese, Mongolian, Korean, Southeast Asian, and Pacific Islander populations.

4. **Native American:** Peoples that migrated from Asia to populate North, Central, and South America.

FIVE FREAKY FACTS ABOUT...
MUMMIES

- The average Egyptian mummy contains more than 20 layers of cloth that, laid end-to-end, would be more than four football fields long.

- In 1974 a scientist discovered that the mummy of Pharaoh Ramesses II, more than 3,000 years old, had a fungal infection. So it was sent to France for treatment, complete with an Egyptian passport describing its occupation as "King, deceased."

- What's the quickest way to tell if an Egyptian mummy still has its brains? Shake the skull—if it rattles, the brain is still in there.

- The Egyptians mummified bulls, cats, baboons, birds, crocodiles, fish, scorpions, insects, and wild dogs. One tomb had more than one million mummified birds.

- Some mummies have been discovered in coffins containing chicken bones. Some scientists believe the bones have special religious meaning, but others theorize that the bones are actually leftover garbage from an embalmer's lunch.

TRIMETHYLAMINURIA

People with this metabolic disorder are unable to produce an enzyme used to break down trimethylamine, a compound found in many foods. That means trimethylamine builds up in the body, until there is so much that it is finally emitted through urine, sweat, and breathing. And that's bad...because trimethylamine smells like stinky fish. In fact, it's the very chemical compound that makes stinky fish smell like stinky fish in the first place. People with this disorder can smell so strongly that— as you can imagine—it can make social situations difficult. Trimethylaminuria is a genetic disease, and there is no cure, though the strength of the fishy odor can be reduced through diet.

EXTRA: The gene mutation that causes trimethylaminuria was discovered in 1997, but the disease has been around for a long time. The earliest known mention of it may come from the ancient Sanskrit epic the *Mahabharata*, a collection of Indian folk tales more than 2,000 years old. One of those tales includes a maiden who "grew to be comely and fair, but a fishy odor ever clung to her."

Inspired by Fiction

Scientists proved they have a sense of humor when they named these four species after fictional characters.

SPONGIFORMA SQUAREPANTSII. San Francisco State University researcher Dennis Desjardin gave this mushroom species its name in 2011 because the mushroom not only resembled a sea sponge, but when viewed under a scanning-electron microscope, it looked like the ocean floor where TV cartoon character SpongeBob SquarePants lives.

ERECHTHIAS BEEBLEBROXI. This moth species has a bump near its head that gives it the appearance of having two heads, just like Zaphod Beeblebrox—the two-headed character in Douglas Adams's *The Hitchhiker's Guide to the Galaxy.*

ERIOVIXIA GRYFFINDORI. In 2016 researchers Javed Ahmed and Rajashree Khalap, both fans of J. K. Rowling's Harry Potter books, discovered a small spider in India whose shape resembles the magic Sorting Hat that once belonged to the wizard Gryffindor in the series—and named it after him.

LEPIDOPA LUCIAE. American crustacean biologist Christopher Boyko named this marine crab in 2002. At first he wanted to name it for Charles Schulz, the creator of the *Peanuts* comic strip. But Schulz's wife suggested he name it after *Peanuts* character Lucy instead...because she's known for being crabby.

SPACE WOMAN

Millie Hughes-Fulford was born in Mineral Wells,
Texas, in 1945; her dad ran a local grocery store.
But her passion lay in science, not commerce, and
she was inspired by great female scientists like
Marie Curie. By 16, she could run the store on her
own, but longed to go to college. So she went off to
earn a PhD in radiation biochemistry from Texas
Woman's University. As if becoming a prominent
medical researcher and molecular biologist weren't
enough, Fulford joined the NASA astronaut
program as a payload specialist aboard the STS-
40 Spacelab Life Sciences (SLS1). This nine-day
mission onboard the space shuttle *Columbia* lifted
off on June 5, 1991. During the mission, Fulford
conducted scientific experiments in space that
yielded critical information about bone-cell growth,
osteoporosis, and cancer.

SPOT OF MYSTERY

In the early 20th century, Gold Hill, Oregon, was a small town with little gold left to mine. Sometime in the 1910s, after a rainstorm, a former gold processing office (essentially a small shack) slid off of its foundation and came to rest down a hillside. People went inside and found a place that didn't make any sense: rocks rolled uphill, and people were able to stand on the walls.

In 1930 that shack—now called the Oregon Vortex House of Mystery—opened as a tourist attraction where visitors could see the strangeness for themselves:

- Brooms stand on end.
- People get taller or shorter as they move through the house.
- Few animals are willing to enter the area.

BRING IN THE SCIENTIST

Scottish engineer John Litster came to the Vortex in 1929 to study all the strange goings-on. He reportedly even discussed what he called "abnormalities" in the area's magnetic field with Albert Einstein. Litster thought there was something extraordinary happening, but he never shared his findings with anyone. To this day, no one really knows what he came up with. Legend has it that, before he died in 1959, he burned all his notes, but supposedly he'd written, "The world isn't yet ready for what goes on here."

SO WHAT'S REALLY GOING ON?

Over the years, people came up with various theories:

1. When the house fell, it moved so fast that it ripped a hole in the earth and created a "gravitational anomaly" where "high-velocity soft electrons" exit the earth. (There are no such things as "gravitational anomalies" or "high-velocity soft electrons.")

2. There's a giant underground magnet that's causing the strange activity.

3. The weirdness could be caused by magnetic rocks in the area. (There is no evidence of iron or other magnetic rocks in the ground near Gold Hill.)

4. High concentrations of volcanic rocks cause the strange events. (Volcanic rocks can't alter gravity.)

THE EXPLANATION?

The debate has raged for years. True believers point to magnetic abnormalities, or even the supernatural. But most think the tricks at the Vortex are optical illusions. The House of Mystery is tilted at a strange incline. The floor, walls, and ceiling are built at sloping angles to trick people into thinking everything looks distorted.

THE SCIENCE BEHIND TOYS

SLIME

Toy slime—sold by Mattel from the 1970s to the 1990s—is what's called a non-Newtonian fluid. That means it changes density depending on how much pressure is applied. Use a light touch, and a non-Newtonian fluid feels as thin as water. Press hard, and it feels thick. In slime, it's the ratio of polymer to gelling agent that makes it a non-Newtonian fluid. Usually, that's a 5-to-1 ratio of polyvinyl alcohol (the polymer) to borax (the gelling agent); the rest is water, fragrance, and coloring.

MAGNA DOODLE

What lets you draw on a Magna Doodle time and time again? There's a layer of honeycombed plastic under the top screen, and each honeycomb cell contains thickened water and magnetic particles. When the magnet on the pen is drawn over the screen, it pulls the particles to the surface, and the water solution is thick enough that they can't float back down. (The solution is also colored white so that the particles are more visible.) When the picture is erased, a magnetic bar along the bottom pulls all the particles back down; the water's thickness keeps them from floating up until the pen is used again.

WHO'S YOUR DADDY?

As geneticists have been explaining for decades, when a mommy and daddy love each other very much, the daddy gives the mommy the genetic material required to make a baby. (This also sometimes happens if they don't love each other.) In male lineages (father to son), the Y chromosome's DNA sequencing remains unchanged for the most part, save a few mutations across generations.

In 2003 scientists studied blood samples from across Asia and found a peculiar trend: a common Y chromosome surfaced across 16 diverse populations. Using deductive reasoning and generational mathematics, they estimated that this specific Y chromosome became embedded into genetic codes approximately 1,000 years ago. Impressive, given that this was during a time when there were no cars or trains and the disbursement covered a wide range of unforgiving terrains.

But who was this biological Casanova? The geneticists turned to the history of the region to investigate further. The timing aligned with the expansion of the Mongolian

Empire in the 12th century, led by none other than the infamous Genghis Khan. Legend has it that the conquering Mongolians saved the most beautiful women for their Universal Leader (as his name translates). So it appears he was universally busy.

Because Genghis Khan's final resting place is unknown, there is no direct DNA to confirm this link. But comparison of the modern-day Y chromosome with that of Khan's known sons, many of whom went on to lead countries and have significant legacies themselves, upholds the theory. For comparison, a man who fathered one child during Genghis Khan's time would have about 800 descendants today based on average population growth. Experts project that Genghis Khan has at least 16 million male descendants, 0.5% of the male population, as of today. Talk about a lasting legacy!

 3 FACTS ABOUT LIGHTNING

1. According to a University of Michigan study, men are six times more likely to be struck by lightning than women are.
2. Lightning can heat the air around it to temperatures of more than 50,000°F.
3. The odds of being killed by lightning are about the same as being killed falling out of bed.

Magic Chip

Have you seen the chip that gets taken out of Jason Bourne's hip in *The Bourne Identity*? How about the one that's injected into Katniss Everdeen's arm in *The Hunger Games*? Then you know just what this thing looks like. The xNTimplantable NFC (Near Field Communication) chip gets implanted into your hand between your thumb and index finger. Once it's there, the chip emits a low-power radio-frequency signature that can "trigger preprogrammed events." In layman's terms, that means it can open locks, start a car, and unlock a computer or smartphone. All you have to do is wave your hand at the lock or device. But first...you have to either inject it into your hand or find someone else who's willing to do so.

Spooky Action

Quantum mechanics is a strange field. It deals with subatomic particles that exist in two different places at the same time, a phenomenon called *superposition*. Albert Einstein called this "spooky action at a distance," but he couldn't explain how it could occur. In 2015 researchers at Stanford University demonstrated superposition by using a 33-foot-tall chamber, lasers, and "10,000 rubidium atoms cooled to near-absolute zero" to get a single atom to fall from two different spots 1.77 feet apart. But that doesn't explain *why* superposition occurs. Bill Poirier, a Texas Tech University professor of chemistry, put forth a fascinating theory: atoms exist in two places because they are in parallel universes that interact with ours.

His theory, called "Many Interacting Worlds," has been well received—surprising, considering it sounds like the title of a heady science fiction novel. "At a symposium in Vienna in 2013," Poirier boasts, "standing five feet away from a famous Nobel Laureate in physics, I gave my presentation on this work fully expecting criticism. I was surprised when I received none." His math checks out. So get ready to hear a lot more about parallel universes in the near future.

FIVE FREAKY FACTS ABOUT...
NEIL DEGRASSE TYSON

- Before he was an astrophysicist, Neil deGrasse Tyson won a gold medal for Latin ballroom dancing in a national tournament.

- He received hate mail, mostly from children, after he insisted Pluto doesn't meet the qualifications to be considered a planet. (It's a dwarf planet now.)

- When he noticed the constellations in the 1997 movie *Titanic* were wrong, he convinced director James Cameron to fix the starry sky in the 2012 rerelease.

- He granted permission to, and even consulted with, DC Comics to include him on a storyline in which he helps Superman locate the alien planet Krypton.

- He once visited a male strip club to look into becoming an exotic dancer there.

ROSETTA STONE

In 1798 French general Napoléon Bonaparte conquered Egypt. The next year, the French army was building Fort Julien near the Mediterranean port city of Rosetta (now Rashid), and army engineer Pierre-Francois Bouchard discovered a blue-gray *granodiorite* stone, 45 inches high, 29 inches wide, and 11 inches thick, covered in three kinds of writing. At the top were Egyptian hieroglyphics, in the middle was conversational Egyptian (Demotic), and at the bottom was classical Greek. The army sent the stone to the Institut de l'Egypte in Cairo, and it took a team of scholars 23 years to fully decipher it. They deciphered the Greek first and found that it was a decree written in 196 B.C. honoring the Egyptian ruler Ptolemy V. Then they worked backward, translating the Demotic and, ultimately, the hieroglyphics. The discovery—and decoding—of the Rosetta Stone was the first key to understanding Egyptian hieroglyphics. The Rosetta Stone has been on display at the British Museum in London since 1802.

SOVIET POISON TRIALS

Soviet biochemist Grigory Mairanovsky was the head of "Laboratory 1," a super-secret Moscow facility run by the KGB, from 1939 until 1946. His assignment: develop a flavorless, odorless poison that is undetectable in an autopsy. To achieve this, Mairanovsky personally directed experiments on humans—all of them political prisoners. They were given poisons with meals or in drinks, and tested for effects. (They were kept in bare cells and observed through small windows.) If the poison failed to kill them, the prisoner would be nursed back to health to await another round. Records show that more than 100 people were tested in this way. Not one survived, and many died agonizing deaths. It's rumored that Mairanovsky did succeed in developing a poison (dubbed "C-2") during his time at Laboratory 1. It could reportedly kill within 15 minutes and was undetectable in an autopsy.

UPDATE: The rumors have never been confirmed.

A lobster's brain is about the size of a ballpoint pen's tip.

Mr. Bell's Assistant

It's one of the most famous moments in the history of invention. It's 1876, and Alexander Graham Bell has been struggling for weeks to get his telephone contraption to work. He yells: "Mr. Watson! Come here! I want to see you!" Thomas Watson, sitting in the next room behind a closed door, hears the scream over his crude telephone receiver. The first telephone call has been made—and the world will never be the same.

Thomas Augustus Watson was an essential contributor to the invention of the telephone. He was only 22 when he got that famous phone call, but Bell might never have made it if it hadn't been for Watson's knowledge of electrical devices, particularly wound-coil electrical devices, which were the key to Bell's big breakthrough.

Watson worked for several years as the Bell Telephone Company's chief repairman, and testified for the company at patent infringement trials. He also invented the telephone booth—his prototype was a tunnel of blankets

used to insulate his voice so his landlady wouldn't complain about the noise. In 1883 he perfected the design using a wood frame, domed top, ventilator, windows, and a desk with a pen and ink.

By the time he was 27, flush with patent royalties, he quit Bell. He started his own machine shop, building marine engines, and later, ships. In Braintree, Massachusetts, he helped with the construction of schools—including the town's first night school and first kindergarten—and often paid teachers from his own pocket. He established the town's first electric plant and streetlights. In 1903 he took up geology at MIT, then traveled to Alaska and California to prospect for precious ores.

In 1915, as part of the launch of transatlantic telephone service, Watson and Bell reenacted their famous conversation, this time with Watson in San Francisco and Bell in New York. In response to Bell's "Mr. Watson! Come here!" Watson replied he would be glad to "but that it would take more than a week."

SALT OF THE EARTH

- It's acceptable to consume about 5,000 mg of salt a day. If you eat more than ¼ cup at once, you'll die.

- Salt is made of two elements—sodium and chlorine—which, if put in your mouth by themselves, will either blow up (sodium) or poison you (chlorine). But merged into a compound—sodium chloride—they change into an essential of life. The salt taste comes from the chlorine—which is also vital for making hydrochloric acid, which digests food in our stomach.

- Scientists once thought the oceans were salty because rivers constantly washed salt out of soil and carried it to sea. But then they found pools of seawater trapped in underground sediments millions of years ago that show the ocean has always been about as salty as it is now.

- There's enough salt in the oceans to cover the world 14 inches deep.

- Salt is the only mineral that can be mined by turning it into a liquid (by pumping water in). Then they pump out the brine and turn it back into a solid by evaporation.

- Salt is hygroscopic, which means it absorbs water. That's why you can't drink seawater; it will dehydrate you.

- Salt is one of the five things the tongue can taste (the others are sweet, sour, bitter, and umami, or savory). Sweet and bitter are inborn, while umami and salty are acquired tastes.

- The hypothalamus at the base of the brain measures sodium and potassium in body fluids. When they get too high (from either not drinking enough water or eating too much salt), it triggers the sensation of thirst.

- When salt is made by vigorous boiling, it forms cubic crystals, but when it's naturally dried, it makes pyramid-shaped crystals. The pyramid-shaped crystals are particularly sought after for kosher use and in fine cooking.

- It takes four gallons of seawater to make a pound of salt.

- For centuries, salt was served in a bowl, not a shaker. It couldn't be shaken, since it absorbs water and sticks together. The Morton Salt Co. changed that in 1910 by covering every grain with chemicals that keep water out—thus its famous slogan, "When it rains, it pours."

MANIMALS!

BIRD–BIRD HYBRIDS

Evan Balaban, a behavioral neuroscientist at McGill University in Montreal, has produced *bird-bird* hybrids: He took brain cells from embryonic quails and transferred them into the brains of embryonic chickens. When the chickens later hatched and grew up, they didn't "cluck" or "cock-a-doodle-doo" like normal chickens...they trilled like quail. And they bobbed their heads just like quail do. Balaban said the work upended the long-held belief that these behaviors are learned, showing conclusively that they are not only hardwired—but that they can be transferred to entirely different species.

MORE MOVIE MAD SCIENTISTS

DR. JEKYLL AND MR. HYDE

Long before Anthony Hopkins got an Oscar for playing a doctor gone bad in *Silence of the Lambs,* Fredric March copped one in 1931 for this gem. You know how it goes: Mild-mannered doctor by day becomes an evil criminal by night. This one's been remade a few times (including as a stoner comedy in the early 1980s, for which karmic punishment will certainly apply), but the Fredric March version is still the best.

THE ADVENTURES OF BUCKAROO BANZAI ACROSS THE 8TH DIMENSION!

The mad scientist is Dr. Emilio Lizardo (John Lithgow), who went looking for trouble in the eighth dimension and found it when some goopy-looking alien took over his skull. Now he needs to get back to where he once belonged, and the only thing stopping him is Buckaroo Banzai (Peter Weller): scientist, rock 'n' roll star, and cultural icon. A true cult favorite among the brainy and socially maladapted. While it is a little obscure, it starts making twisted sense after the fifth or sixth viewing.

"REAL" APHRODISIACS

Traditionally, scientists have dismissed aphrodisiacs as frauds. But new research into medicinal herbs and pheromones (chemical messengers) has produced some interesting results. Experts now believe that some aphrodisiacs may really work.

YOHIMBE: For centuries, the bark of the West African yohimbe tree has been thought to produce passion in African men. Research has found that the chemical yohimbine in the tree can in fact excite men by increasing blood flow. The drug was approved by the FDA 10 years ago as a prescription treatment for impotence.

OYSTERS: Traditionally considered an aphrodisiac because of their association with the sea and their resemblance to female sex organs. However, now we also know that they're very rich in zinc—a mineral necessary to male sexual health. A man deficient in zinc is at high risk for infertility and loss of libido.

CHOCOLATE: Contains PEA, a neurotransmitter that is a natural form of the stimulant amphetamine. It has been shown that either love or lust increases the level of PEA in the bloodstream and that with heartbreak, the levels drop dramatically.

More "Real" Aphrodisiacs

CAFFEINE: Research has shown that coffee drinkers are more sexually active than non-drinkers, but no one's sure if that's because of something in the caffeine, or just because it keeps people awake, and therefore interested, after bedtime.

DHEA: This hormone has been called the "natural aphrodisiac" by doctors. It's been shown in studies that blood levels of DHEA predict sexual thoughts and desire. DHEA became a food-supplement fad when it was hyped in the media as a way to increase energy and maybe even prevent cancer or heart disease (as well as boosting the libido).

CINNAMON: According to Dr. Alan Hirsch, director of the Smell and Taste Research Foundation, the aroma of cinnamon has the ability to arouse lust. As reported in *Psychology Today*, "Hirsch fitted male medical students with gauges that detected their excitement level, and then exposed them to dozens of fragrances. The only one that got a rise was the smell of hot cinnamon buns."

ANDROSTENONE: This is a pheromone. Scientists conducting research with animals found that androstenone produced by boars had a very positive effect on the sexual receptivity of sows. Androstenone is also found in human sweat.

WHAT A SHOCKER

Shenandoah National Park ranger Roy Sullivan claims he was struck by lightning an unbelievable seven times between 1942 and 1977!

STRIKE #1 (1942): While on duty in one of Shenandoah's fire lookout towers, Sullivan took his first hit. The lightning bolt hit his leg, and he lost a big toenail.

STRIKE #2 (1969): This time, he was driving on a country road. The lightning hit his truck, knocked him out, and singed off his eyebrows.

STRIKE #3 (1970): People started calling Sullivan the "human lightning rod" after the third strike, which injured his shoulder.

STRIKE #4 (1972): He took this hit while on duty at one of Shenandoah's ranger stations. The lightning set his hair on fire, so Sullivan started carrying a bucket of water around with him—just in case he needed to put out a blaze.

STRIKE #5 (1973): This one also set his hair on fire (thank goodness for that water bucket!). And it knocked him out of his car and blew off one of his shoes.

STRIKE #6 (1974): Lightning hit Sullivan at a park campground, and he hurt his ankle.

STRIKE #7 (1977): This might have been his most dangerous strike. The lightning hit him while he was fishing and burned his stomach and chest, requiring a hospital stay. (He recovered.)

HONORABLE MENTION: Sullivan's wife was also hit by lightning once while she and Roy were hanging up clothes on a line in their backyard.

SNOWBOARDING SCIENCE

If you haven't already read "Skateboard Science,"
go to page 33 first. Then cruise on back!

By the 1980s snowboarders were "shredding" the slopes
(sliding downhill with their feet strapped to a board);
they'd adapted the skills of skateboarding to snow. Like
skateboarders, they rode either regular (with their left foot in
front) or goofy foot (with their right foot in front). They even
adapted the half-pipe, picking up enough momentum in a
high curving trench packed with snow so they could slide
up over the top of the lip and catch some awesome—but
cold—air.

Snowboarders also take advantage of the forces of friction,
gravity, acceleration, and momentum. A board speeds downhill
pulled, of course, by gravity, but it also melts the snow as it
goes, so that it actually zips along on a film of water.

Like all other types of board riders, snowboarders position
their critical mass and exploit the board's rotational motion
to stay balanced. One side of the board will have more contact
with the snow than the other. A rider keeps his center of
gravity over whichever edge of the board is in contact with the
snow (the riding edge). To end a ride, a snowboarder turns
uphill so that the force of friction and the force of gravity drag
on the momentum and slow the board to a stop.

Schrödinger Simplified

If you happen to find yourself in a conversation about quantum theory, chances are that Schrödinger's cat will pop up...which is ironic because its original intention was to illustrate just how absurd quantum theory is.

In 1935 Erwin Schrödinger and fellow Austrian physicist Albert Einstein were discussing *quantum superposition,* the theory that one subatomic particle can exist in two places at the same time (see page 172 for more). That's when Schrödinger came up with his famous "thought experiment." It goes like this: if you seal a cat inside a box along with one vial of hydrocyanic acid, there is a 50-50 chance that an atom will decay. If it does, the acid will release a radioactive substance that will kill the cat.*

Quantum theory says that, because there is no way to know if the cat is alive or dead inside the sealed box, then it is both alive *and* dead—existing in two states at the same time. But the only way to know the true state of the cat is to open the box. This is called the *observer's paradox*: "The observation or measurement itself affects an outcome, so that the outcome as such does not exist unless the measurement is made." Schrödinger was trying to point out that scientists can say all they want that one particle can exist in two places at once, but as soon as you observe it, it's only ever one particle.

However, by the 1950s, the quantum superposition theory had gained such traction in the scientific world that Schrödinger himself wanted to "disown the cat." But his friend Einstein wrote him a letter of reassurance:

> You are the only contemporary physicist...who sees that one cannot get around the assumption of reality—if only one is honest. Most of them simply do not see what sort of risky game they are playing with reality—reality as something independent of what is experimentally established...Nobody really doubts that the presence or absence of the cat is something independent of the act of observation.

* No cats were harmed in the writing of this article.

DUMB

PREDICTIONS

"That virus is a pussycat."

—Dr. Peter Duesberg,
molecular biology professor
at U.C. Berkeley, 1988
(commenting on HIV)

WHIPWORM THERAPY

Whipworms are a type of helminth—a worm classified as a parasite. Long, thin, and pale, whipworms thrive in places that lack proper sanitation. Their microscopic eggs are accidentally eaten, and then...they hatch. The newly hatched worms migrate to the intestines, attach themselves to the walls, and chow down. Whipworms can cause everything from intense stomach distress to retardation in children. In developed countries, better sanitation has largely wiped out whipworms.

RETURN OF THE WHIPWORM: However, whipworms may now be needed to treat disease. Some doctors now believe it's possible for people to be too clean. (Huh?) Our immune systems are made to attack outside invaders like bacteria, germs, and parasites. But when invaders are in short supply, a person's immune system can misfire. The result: an autoimmune disease, which happens when the immune system attacks a person's own body instead of an invader. People infected with whipworms rarely suffer from autoimmune diseases. Doctors wanted to see if patients' immune systems could be reeducated to attack whipworms and leave the body alone. Patients slurped down a dose of several hundred whipworm eggs in salty liquid and then let them grow. The little wrigglers did such a good job they may soon be available by prescription.

The Robots Take Over

DROPLET takes the guesswork out of watering your plants; this small robot attaches to a watering hose and is connected to the Internet, so it will hold off on watering if there is a good chance of rain in the forecast. It can be programmed to supply different amounts of water to each plant, so you won't have to rely on your absentminded neighbors to water your plants when you're on vacation.

KILOBOTS are the size of a quarter, which doesn't sound like much—but imagine 1,024 tiny robots working together in a swarm like an army of ants, forming various shapes and adjusting on the fly to explore dangerous places or carry out complex tasks. Developed at Harvard, these tiny bots could one day be deployed for rescue missions...or the technology could grow exponentially and one day we'll see millions of kilobots rise up against their human oppressors.

EARTH'S CARETAKERS

To celebrate its 10th anniversary, the Environment Agency in Britain invited a blue-ribbon panel of ecologists to list its 100 greatest "eco-heroes" of all time. Rachel Carson, American scientist and author of *Silent Spring* (1962), was number one on the list. The panelists dubbed her "the patron saint of the green movement." Also included were the ancient Greek philosopher Aristotle; Wangari Maathai, known as Africa's "tree woman"; Siddartha Gautama Buddha, a meditation master who lived in the sixth century B.C.; and St. Francis of Assisi, Catholic patron saint of animals and ecology.

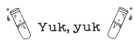
Yuk, yuk

Q: What does a subatomic duck say?

A: "Quark!"

Albert B.,
The Lab Rat

John B. Watson was a behavioral psychologist at Johns
Hopkins University when he performed an experiment that
still has people guessing at the results. Watson believed
that people are the way they are based solely on their
environment. In his view, environmental conditioning is
what makes a baby brave or fearful; it can also influence
him or her to grow up to become virtually anything from
a biologist to a burglar, regardless of natural talents or
tendencies. In 1920 he set out to prove his theory.

The unfortunate subject of his experiment was a nine-month-old known only as "Little Albert" or "Albert B."

Albert showed no fear when he was allowed to play with a white rat—until Watson's experiment. Every time Albert touched the rat, Watson or his assistant made a loud clanging noise that scared the baby and made him cry. Albert learned to associate the scary sound with the rat, and eventually, even without the sound, he was afraid of the rat. This fear carried over to any furry creature that reminded Albert of the rat. Success! Watson had just conditioned a phobic toddler.

It's still not known if Albert's mother (a single woman employed as a wet nurse) knew every single detail of the experiment, or where she took her little boy, but take him away she did—before Watson could reverse his conditioning—if he planned to.

Watson's findings created an uproar: How could he have preyed on an innocent child—and would Little Albert ever recover from his ordeal? Decades later, researchers discovered that Little Albert's real name was Douglas Merritte. But we'll never know if Douglas continued to live in fear of furry animals—he died of an illness when he was six years old.

VANISHING LAKES MYSTERY

In 2015 oceanographers from MIT announced that they had solved a decades-old mystery: How did the large lakes that form during the summer on top of Greenland's ice sheets vanish in less than 24 hours? And where did the water go?

• The appearance of summertime lakes, known as *supraglacial* or *meltwater* lakes, has increased in recent years, as has their size. Thousands of them, some covering up to a few square miles, show up each summer. And many of them do the baffling disappearing act. Where could so much water go so quickly?

• The answer came when oceanographers placed seismic instruments around Greenland's North Lake in 2011, 2012, and 2013...and waited. The lake covered 2.2 square miles and was as deep as 40 feet in some places, giving it in the neighborhood of 11.6 billion gallons of water. The team was able to record 90% of the lake's water disappearing in 90 minutes. It had drained through massive cracks that the weight of the water had made in the ice. That explanation

had been considered before, but such cracks would have to reach the bottom of the ice sheet to allow all the water to disappear, and that had been deemed impossible. It wasn't. The cracks opened up to a depth of more than 3,200 feet—all the way to where the glacier sits on bedrock and slowly makes its way to the ocean. That allowed the millions of tons of water to drain almost at once—with more force than Niagara Falls—to the bedrock underneath the ice, which actually caused a huge section of the ice to rise some 20 feet as the water surged beneath it.

"All of the biggest technological inventions created by man—the airplane, the automobile, the computer—says little about his intelligence, but speaks volumes about his laziness."

–Mark Kennedy (politician)

EXPENDABLE ORGANS

TONSILS

Every year about 530,000 kids under the age of 15 have their tonsils removed. It's one of the most common organ removals, and probably the oldest. A Hindi medical guide from about 1000 B.C. holds the first known instructions for doing a tonsillectomy: "When troublesome, they are to be seized between the blades of a forceps, drawn forward, and with a semicircular knife, a third of the swelled part is removed." Roman doctors in the first century A.D. also debated whether full or partial removal was best.

WHAT THEY'RE FOR: Tonsils are specialized lymph nodes that help filter bacteria and viruses out of the blood.

WHY WE CAN DO WITHOUT THEM: When they get infected, tonsils can cause sore throats and apnea, but if they're removed, other lymph nodes take over their function.

SIDE EFFECTS OF REMOVAL? About one in 15,000 patients dies from bleeding, reactions to anesthesia, or airway obstruction. In a few patients, throat problems get worse. A review of 7,765 research papers in 2009 found that the positive effects in kids were generally modest and short-lived.

ADENOIDS

In a way, the adenoids are sort of the tonsils of the nose—they rest in the base of the nose and trap germs. When infected, they can impair nose breathing and increase chronic infections and earaches. They are often removed at the same time as the tonsils.

WHAT THEY'RE FOR: Filtering bacteria and viruses.

WHY WE CAN DO WITHOUT THEM: For babies under one year of age, the adenoids are an important part of the immune system, but after that age, they become increasingly irrelevant.

SIDE EFFECTS OF REMOVAL? See tonsillectomy on the previous page. Recent studies have also called into question the effectiveness of adenoid removal for preventing respiratory infections. Even for the condition most helped by an adenoidectomy—snoring and near-suffocation caused by sleep apnea—some patients saw no improvement afterward.

MYTHUNDERSTANDINGS

MORSE CODE

MYTH: The dot-dash code for telegraphs is named after the code's creator, Samuel Morse.

THE TRUTH: The code *is* named after Morse—actually, he named it after himself. But many historians think his collaborator, Alfred Vail, actually created it. Morse's original notes from 1832 suggest that he was planning a code that assigned each word in the dictionary a number. But six years later, someone came up with an alphabet code, using dots and dashes to signify letters. An apprentice, William Baxter, said it was Vail's innovation. But Morse insisted he'd done it, and history books simply take his word for it.

IT'S RAINING AMPHIBIANS!

Once in a great while, people are forced to seek cover from a storm that drops not just rain, but...*live animals.* It happened in England in the 1800s, when jellyfish fell from the clouds. In the 1930s, frogs followed suit. In 2010 a remote town in Australia experienced showers of hundreds of spangled perch two days in a row. The kicker? The town was more than 200 miles from the nearest body of water. This phenomenon is truly a mystery. Some scientists theorize that massive evaporation, a strong updraft, or a tornado picks up aquatic life along with water, blows it 70,000 feet high, and carries it many miles before the winds die down and the animals fall with the rain. However, this doesn't explain the situations in which only a single species rains at once (like the perch in Australia), instead of a mixture of different animals. People in Honduras have a legend to explain this phenomena.

Every summer in Yoro, Honduras, fish rain down during thunderstorms. The Hondurans believe it's an answered prayer. Long ago, a Catholic priest prayed for food for the starving natives. Supposedly, that's when the fish rains began. Today, some of the fish fall into waterways and swim off. What do villagers do with the rest? Cook and eat them, of course. Waste not, want not.

ribbit

Pop (Culture) Science

These science facts sound like they're
straight out of a movie...

A WALKING TANK

Looking like something that Tony "Iron Man" Stark might
wear, the TALOS, or Tactical Assault Light Operator Suit,
is the U.S. Army's newest toy. The suit's liquid body armor
will transform into a hard surface when struck by a bullet
or shrapnel. It will even have sensors that will monitor the
soldier's vital signs. The army hopes to have the suits in the
field by 2018. (No flight capabilities are included. Yet.)

FACE THE FUTURE

Remember that scene in *Minority Report* when Tom Cruise
walks into a store and a computer knows who he is? That's
a real thing now. Retail giant Tesco has installed facial
recognition scanners at its gas stations around the UK:
When a customer pays, his or her face is scanned to identify
age and gender. Then an advertisement plays, specifically
targeted to sell to the customer's demographic. Expect to
see the scanners everywhere in the near future...unless they
see you first.

Green City:
VANCOUVER

POPULATION: 630,000

HOW GREEN IS IT? Often called the greenest city in Canada, Vancouver has more than 200 parks in a region that's surrounded by spectacular beaches, forests, and mountains. The city leads the world in the production of hydropower, which supplies 90 percent of its electricity. And one of Vancouver's most famous innovations is the use of solar-powered trash-compactor bins on public sidewalks: The bins can hold five times the amount of conventional trash cans, so they need to be emptied only once a week instead of every night, which saves on the need to use the city's gas-powered fleet of garbage trucks.

Vancouver has also been adding new streetcar lines and bike lanes, and it has constructed nearly 250 miles of "greenways," special corridors for pedestrians and cyclists that connect parks, nature reserves, historic sites, neighborhoods, and shopping areas. And 40 percent of commuter and tourist day trips in Vancouver involve walking, biking, or using public transportation.

STRANGE SCIENCE

DOCTOR STRANGE, LOVE

BUSTED

A German plastic surgeon cheated out of payments from four women upon whom he had performed breast enhancement surgery took his complaint to police...along with photos of the women's enhanced breasts. "They registered under fake names," Dr. Michael Koenig told reporters. "And then after the operations, they just ran away." Each surgery cost nearly $14,000. Police made posters of the enhanced-breast photos and distributed them to the public; the German newspaper *Bild* even printed one of the shots. "It's probably the most unusual wanted poster ever," *Bild* wrote.

GOTTA HAND IT TO HER

A doctor in New Brunswick, New Jersey, was arrested in September 2006 after stealing a hand from a cadaver and giving it to an exotic dancer as a gift. She kept it in a jar of formaldehyde. It was discovered when police were called to the woman's apartment because of a suicidal roommate. Friends said she had named the hand "Freddy."

DUMB
PREDICTIONS

"There is practically no
chance communications space
satellites will be used to provide
better telephone, telegraph,
television, or radio service
inside the United States."

—T. A. M. CRAVEN, FCC COMMISSIONER, 1961

THE FIVE MAJOR EXTINCTIONS

The National Wildlife Federation estimates that nearly 30,000 species per year—about three per hour—go extinct, many from man-made causes. Many biologists believe this could be the "sixth extinction," the latest in a series of mass disappearances of species and ecosystems. Here are the first five in Earth's past:

1. The Ordovician-Silurian Extinction (440 million years ago) was a result of climate change caused by sudden global cooling, particularly in the tropical oceans. There was little or no life on land at the time, but nearly 85 percent of marine species disappeared.

2. The Late Devonian Extinction (370 million years ago) may have been caused by asteroid collisions that triggered volcanic lava floods, or environmental changes due to tectonic plate movements in the early formation of the continents. Another theory: the "Devonian Plant Hypothesis," which claims that the expansion of plant life on land caused the death of 75 percent of animal species in the oceans.

3. The Permian-Triassic Extinction (250 million years ago) was the deadliest mass extinction. Paleontologists once believed that this took millions of years to occur, but many now think it took between 8,000 and 100,000 years. The impact and explosion of a large meteor might have accelerated the climate change that was already taking place as a result of greenhouse gases released by volcanic lava floods and tectonic plate shifts. In all, 95 percent of existing species were wiped out.

4. The End Triassic Extinction (200 to 215 million years ago) came soon after dinosaurs and mammals first evolved, and is the most mysterious of the five mass extinctions. In addition to the 75 percent of all species that were lost, an unknown number of land-dwelling vertebrates also went extinct. Causes similar to the previous Permian-Triassic Extinction are suspected, including volcanic lava floods, shifting continents, and meteor impacts.

5. The Cretaceous-Tertiary Extinction (65 million years ago) is the most recent and best known of the five mass extinctions. One or several meteor impacts are probably to blame, which caused lava floods that so completely disrupted Earth's ecosystems that many terrestrial and marine species rapidly went extinct. The extinction killed 75 percent of all species, including—most famously—the dinosaurs.

RANDOM ORIGIN

CELL PHONES

AT&T first tested mobile phones for use in Swedish police cars in 1946. To develop the technology in the United States, they needed approval from the FCC—which controls the radio waves. The FCC didn't think mobile phones would work and repeatedly turned down AT&T...until 1968, when AT&T unveiled its plan: offer phone service via many low-powered broadcast towers, each covering a "cell" of a few miles. As the car phone user traveled, calls passed from tower to tower uninterrupted.

Meanwhile, rival Motorola had secretly developed their own mobile phone, only theirs was a handheld model. (AT&T had concentrated on car phones.) In 1973 one of Motorola's engineers, Dr. Martin Cooper, used a prototype to make the first cell phone call—to AT&T, to gloat. But AT&T was the first to get FCC approval, and had a trial cellular network set up in Chicago by 1978. The FCC authorized nationwide commercial cellular service in 1982 and just five years later there were over one million cell phone users in the United States.

HOW TO MAKE A MUMMY

Scientists have yet to unlock all of the secrets of Egyptian mummification, but they have a pretty good idea of how the process worked:

- When a king or other high official died, the embalmers slit open the body and removed nearly all the organs, which they preserved separately in special ceremonial jars. A few of the important organs, like the heart and kidneys, were left in place. The embalmers apparently thought the brain was useless and in most cases brain matter was shredded with small hooks inserted through the nostrils, pulled out through the nose using tiny spoons, and thrown away.

- Next, the embalmers packed the body in oil of cedar (similar to turpentine) and natron, a special mineral with a high salt content. The chemicals slowly dried the body out, a process that took from 40 to 70 days.

- The body was then completely dried out and "preserved," but the process invariably left it shrunken and wrinkled like a prune. The next step was to stuff the mouth, nose, chest cavities, etc. with sawdust, pottery, cloth, and other items to fill it out and make it look more human. In many cases the eyes were removed and replaced with artificial eyes.

- Then the embalmers doused the body with a waterproofing substance similar to tar, which protected the dried body from moisture. In fact, the word *mummy* comes from the Persian word *mumiai*, which means "pitch" or "asphalt," and was originally used to describe the preservatives themselves, not the corpse that had been preserved.

- Finally, the body was carefully wrapped in narrow strips of linen, and a funerary mask resembling the deceased was placed on the head. Afterward it was laid in a large coffin that was carved and painted to look like the deceased, and the coffin was placed in a tomb outfitted with the everyday items that the deceased would need in the afterlife.

> "One, remember to look up at the stars and not down at your feet. Two, never give up work. Work gives you meaning and purpose, and life is empty without it. Three, if you are lucky enough to find love, remember it is there and don't throw it away."
> —**Stephen Hawking**

GOVERNMENT WASTE

HIGH-TECH BUS STOP

In 2013 Arlington County, Virginia, received funding to build a bus stop complete with Wi-Fi, heated benches and sidewalks, and "a wall made of etched glass that opens the rear vista to newly planted landscaping." Too bad the slanted glass roof doesn't do much to keep out rain and snow, or provide shade in the summertime.

Cost to taxpayers: $1 million

SILLY SOLAR PANELS

A federal grant was used to install solar panels on the parking garage at the Manchester-Boston airport. One problem: The reflective panels were blinding the pilots, so 25 percent of the panels had to be removed. But the remaining panels, say officials, will generate "$2 million in savings over 25 years."

Cost to taxpayers: $3.6 million

UNSCIENTIFIC STUDY

Executives from various independent music labels received an all-expenses-paid trip to Brazil in 2013 "to compare the record stores, club districts, and facial expressions of locals at the mention of their bands." While the execs reportedly enjoyed their trip, one of them said he "didn't ink any deals."

Cost to taxpayers: $284,300

MORE DREAM DISCOVERIES

THE SEWING MACHINE

Elias Howe had been trying to invent a practical lockstitch sewing machine for years, but had been unsuccessful. One night in the 1840s, he had a nightmare in which he was captured by primitive tribesmen who were threatening to kill him with their spears. Curiously, all the spears had holes in them at the pointed ends. When Howe woke up, he realized that a needle with a hole at its tip—rather than at the base or middle (which is what he'd been working with)—was the solution to his problem.

THE BENZENE MOLECULE

August Kekule, a German chemistry professor, had been working for some time to solve the structural riddle of the benzene molecule. One night while working late, he fell asleep on a chair and dreamed of atoms dancing before him, forming various patterns and structures. He saw long rows of atoms begin to twist like snakes until one of the snakes seized its own tail and began to whirl in a circle. Kekule woke up "as if by a flash of lightning" and began to work out the meaning of his dream image. His discovery of a closed ring with an atom of carbon and hydrogen at each point of a hexagon revolutionized organic chemistry.

HOCKEY SCIENCE

- **"Fast ice"** is hard, cold, and smooth and makes skating and passing easier. Over the course of a hockey period the ice warms up, becomes softer, and its surface gets rougher—**"slow ice."** The puck starts to bounce a little. Players become a little more careful and try to make a safe play instead of a finesse play. A warm puck stores more energy and bounces higher, which gives the player less control.

- The **puck** is a rubber compound mixed with other materials to give it strength and to make it less elastic. It's frozen in a bucket of ice before a game and between periods to reduce some of the bounce. Freezing keeps the puck lower to the ice, where the action usually is.

- A **hockey stick** has to be just flexible enough to store as much energy as possible, and then release it when needed. (A stick that's too flexible wouldn't store enough energy.)

- A slap shot can send that puck toward the goal at well over 100 mph (160 kph). Three factors are involved:

1] The energy produced by the weight the player transfers to the stick by leaning into it.

2] The "stored elastic energy." In hockey, the shaft bends slightly during the swing—the end can't keep up with the handle. The stick stores this energy and releases it when it hits the puck. The result is a greater launching speed than you could get from a nonflexible stick.

3] The snap—a slight snap of the wrists at the end of the motion—releases the puck from the stick. The snap is crucial. It sets the puck to spinning, which makes it more stable in flight. If it wasn't spinning it would roll side over side, which would make it follow a more erratic path.

- Stick-meets-puck isn't the only source of energy on the floor of a rink. Don't forget about hockey players themselves, who have been clocked at more than 20 mph (32 kph) in a rink that's just 200 feet (61 m) long.

A JIFFY

A *jiffy* is a slang term for a very brief amount of time. Its earliest known use dates to the 1780s, though its exact origin is unknown. It's commonly used loosely, as in, "I'll be back in a jiffy," but can also be a term for quite specific amounts of time.

- In electronics, a jiffy is sometimes used as the name for the time required for one alternating current power cycle, 1/60 of a second.

- In computer science, it's sometimes used to describe a microprocessor's "clock cycle," which isn't an absolute interval of time—it decreases as the microprocessor's speed increases. With modern computers, a jiffy used this way could be measured as parts of nanoseconds (billionths of a second).

- In physics (particularly in quantum physics and often in chemistry), a jiffy is the time taken for light to travel the radius of an electron. It's also used in physics as the name for the amount of time it takes light to travel the width of one *nucleon* (a proton or neutron), which would make it by far the jiffiest of all jiffies.

CLONING JOHN LENNON

Sometime between 1964 and 1968 (she doesn't remember when), housekeeper Dot Jarlett was given an extracted molar by her boss, John Lennon. Lennon asked her to throw it away for him, but then jokingly suggested she keep the tooth instead and give it to her daughter, a huge Beatles fan.

In 2011 Omega Auction House acquired it from that lucky Beatles fan and sold it at auction. They expected it to sell for about 9,000 pounds, or around $16,000, which would have been an absurd amount to pay for an old tooth. Omega didn't get that amount for it—it brought in nearly double, 19,500 pounds, or around $31,200.

The buyer was a Canadian dentist named Dr. Michael Zuk. Why'd he buy it? Zuk describes himself as a huge Beatles fan, but that's an understatement considering what he wants to do with the tooth. He is prepared to spend however much it takes to extract Lennon's DNA from the tooth, "fully sequence" it, and then make a clone of John Lennon. Two things are holding back Zuk's plans: cloning science is to the point where a cat or a sheep can be cloned, but not a human. Also, the tooth is so old and so fragile that it was too brittle to be subjected to DNA extraction tests. Zuk is confident that neither of these factors will matter much in the near future. "With researchers working on ways to clone mammoths, the same technology certainly

could make human cloning a reality," Zuk told reporters, referring to a dubious report by Russian scientists.

CLONING AROUND

In the meantime, Zuk allowed his sister to break off a piece of the tooth (presumably one without much precious DNA in it) and use it in an art project—a clay sculpture of John Lennon. The sculpture toured England to raise awareness of mouth cancer.

As for the future Lennon clone, Zuk plans to raise him like a son. He says he'd make him fully aware of his legacy ("Guitar lessons wouldn't hurt anyone, right?"), but introduce some changes, too. "He would still be his exact duplicate, but you know, hopefully keep him away from drugs and cigarettes," Zuk told England's Channel 4.

FIVE FREAKY FACTS ABOUT...
ECOLOGY

- The word *ecology* means "study of the house," from the Greek *eco* for "house" or "environment."

- In 2003 Alaska's Iditarod dogsled race had to be moved 300 miles north—the usual location wasn't cold enough (because of global warming).

- Rain forests cover only 6% of Earth's surface, but they contain more than half the plant and animal species on the planet.

- Almost 70% of Earth's surface is water, but only 1% is usable: 97% is in the ocean and 2% is frozen.

- A group of NASA engineers and American astronomers believe that moving Earth into a new orbit would solve the problem of global warming—or at least add another 6 billion years to its life.

Your Fantastic Feat...Er, Feet

On a mile run, your amazing feet endure about 1,500 heel strikes at a force over two times the body's weight. To a climber, they're grippers and levers. To a skater, they're accelerators, steering mechanisms, brakes, and shock absorbers. To a high jumper, they're levers and launching pads. In most other sports, feet are the literal foundation of performance as they balance, support, and propel an athlete.

Our feet are mobile miracles made up of 26 bones, 33 joints, and 112 ligaments, not to mention the nerves, blood vessels, and tendons that combine to form your personal transportation network. Your feet have three bony arches: a tall one along the inner edge of the foot, a smaller arch on the outer edge, and the curve that runs the width of the foot between the ball and heel. Together they form an arched vault that not only distributes your weight, but is also flexible enough to help you move.

The ligaments that bind the bones of your arch are elastic, so they can flatten out, then spring back to shape. When you take a step, your foot rolls outward and your arch flattens and stiffens into a lever to push your foot off the ground. Then your arch springs back to a curve with an added bounce that propels you along. When you set your foot down, your arch rolls outward and becomes flexible to absorb impact. With every step your foot propels you, stabilizes you, and absorbs shock—all while supporting your weight.

Bad Movie Science

The moral: When movie characters use scientific-sounding words, remember that some Hollywood screenwriter probably made that stuff up.

THE DAY AFTER TOMORROW (2004)

PREMISE: The Gulf Stream, an Atlantic ocean current that helps regulate Earth's temperature, has become so affected by global warming that it essentially stops. The ocean suddenly rises and massive icy tidal waves flood New York City. Within days, North America is a frozen wasteland.

BAD SCIENCE: Global warming can have a detrimental effect on the oceans, but it can't stop the Gulf Stream that fast. Even if it could, in order for New York City to flood like it did in the movie, the entire continent of Antarctica would have to melt. For *that* to happen, all of the sunlight that hits Earth would have to be collectively beamed at the South Pole...for three years.

THE MATRIX (1999)

PREMISE: After the machines take over the world, the human resistance "scorches the sky" to block out the machines' power supply—sunlight. So the machines use the humans for power, keeping them alive in a vegetative state while subjecting their brains to a life simulation. The machines "liquefy the dead so they can be fed intravenously to the living."

BAD SCIENCE: Neither the machines nor the humans know much about sustainable energy production. Blocking out the Sun would just destroy Earth's biosphere; the machines could easily build solar panels in space to get all the power they need. Second, human energy is inefficient—only about 35% of the energy from food converts to mechanical energy. And feeding humans to humans can lead to a disease called kuru, which causes insanity—and would screw up the simulation.

WATERWORLD (1995)

PREMISE: The surface of Earth has been completely covered in water. In one scene, the Mariner (Kevin Costner) swims around an abandoned underwater city that's revealed to be none other than Denver, Colorado, once known as the "Mile High City."

BAD SCIENCE: If the temperature of Earth increased 8°F, sea levels would rise by three feet due to melting polar ice caps, which would be ecologically catastrophic. But sea levels could never rise to the point where Denver was completely submerged—the city's elevation is 5,280 feet. If all the world's ice melted, the ocean would rise 250 feet, submerging many coastal cities, but not Denver.

EINSTEIN'S BLOUSE

Another patent from our
"Dressed By Geniuses" files.

SCIENTIST: Albert Einstein

PATENT NO. US D101756 S: "A new, original, and ornamental blouse"

STORY: A former patent clerk, Einstein was familiar enough with the process to get his own good idea patented in 1936. "The design is characterized by the side openings A-A which also serve as arm holes; a central back panel extends from the yoke to the waistband as indicated at B." According to PatentYogi.com, "It's an expandable suit jacket that has two sets of buttons, one for skinny Albert and one for hefty Albert."

STRANGE SCIENCE

"UNIT 731" EXPERIMENTS

In 1937, during the Second Sino-Japanese War, the Japanese government built an enormous military complex in the puppet state of Manchukuo, in what is now northeast China. Called Unit 731, the facility was headed by General Shiro Ishii, the Japanese army's chief medical officer. Over the course of the following eight years, Ishii directed hundreds of doctors in an unimaginable nightmare of experiments on humans, mostly Chinese and Korean prisoners. This included exposure to biological and chemical warfare agents (such as plague, cholera, and mustard gas), unnecessary amputations, and surgery without painkillers. The experiments were done in the name of medical research, but many had no discernible medical purpose whatsoever. With military defeat in sight, in 1945 General Ishii ordered the executions of all remaining prisoners and fled back to Japan. During his time as the head of Unit 731, more than 10,000 people were experimented upon; roughly 3,000 of them died in the process. Ishii was arrested by U.S. occupation authorities in 1945. He was granted immunity in exchange for information about Unit 731 and received no punishment for his crimes.
UPDATE: Ishii died at home in 1959 at the age of 67.

JULES VERNE, FUTURIST

Our story introducing the futurists is back on page 9.

In 1828, when French writer Jules Verne was born, ocean voyages took months, and there were hardly any railroad tracks. Three decades later, steam-powered ships and locomotives were taking people across oceans and continents in only a week. Knowing that the rate of change was increasing, in 1863 Verne attempted to track it in a book called *Paris in the 20th Century*. Among Verne's predictions for the 1960s: glass skyscrapers, high-speed trains, gas-powered cars, air-conditioned houses, fax machines, and corner stores. His publisher rejected the manuscript as being too "far-fetched."

Verne's next novel, *From the Earth to the Moon*, was a pioneering work of both science fiction and foresight. The plot: three wealthy men finance a trip to the Moon. Their ship was launched from a cannon, so Verne got that part wrong, but he was close to the mark on other details—including the rocket's escape velocity, the Florida launch site (where NASA missions would take place a century later), the three-man crew, and the splashdown in the Pacific. Even more uncanny, Verne's Moon trip cost $5,446,675 ($12 billion in 1969 money). Cost of the actual Moon mission: $14.4 billion.

STRANGE MEDICAL CONDITION

SUBJECT: A 19-year-old Iranian man

CONDITION: "Hairy eyeball"

STORY: In 2013 researchers from Iran's Tabriz University of Medical Sciences reported that they had treated a man who was born with a tiny whitish growth on his right eyeball. For many years it hadn't bothered him, nor had it affected his vision. But it had grown in size, reaching about a quarter-inch in diameter, and it had started to become annoying— especially after several black hairs started to grow out of it. Luckily, those hairs actually helped doctors diagnose the growth: it was a limbal dermoid, a very rare type of tumor that has the bizarre characteristics of being able to grow things like hair, cartilage, bone—and even sweat glands. More good luck: such tumors aren't cancerous. Doctors successfully removed the tumor from the man's eye without causing any lasting damage, and the man resumed his life...minus one hairy eyeball.

FROZEN IN TIME

In Petrified Forest National Park, prehistoric fossils and remnants of ancient civilizations reveal what life was like thousands—even millions—of years ago.

In prehistoric times, the area of northeastern Arizona where the Petrified Forest is located was closer to the equator and was not a desert. It was a floodplain, swollen with streams and rivers. Cycads, horsetails, and ferns dominated the landscape. Coniferous trees were plentiful and large—as much as 200 feet tall and nine feet in diameter. When those trees fell, rivers carried them away, and before they could decompose, some were buried under clay, mud, sand, and volcanic ash. Gradually, minerals in the water leeched into the wood, filling the cracks and crevices of the logs and forming the vivid fossilized logs we have today. Different minerals, of course, made different colors: quartz produced white; manganese oxides made blue, purple, black, and brown; and iron oxides turned the wood yellow, orange, and red. As the trees fossilized over hundreds of millions of years, they turned to stone, and today, many of the logs lie where they fell eons ago. They're better preserved and more colorful than those found anywhere else in the world.

Petrified Forest also includes part of the Painted Desert. It's 7,500 square miles of cliffs, hills, and hardened sand dunes that are "painted" with bright red, green, and yellow bands; many of the rocks also become red, purple, or blue at sunrise and sunset. The colors are the result of mineral deposits left behind by fossilized trees and animals and shaped by wind and water. As the cliffs erode, the fossils and minerals are exposed, changing the desert's colors. Southwest historian Charles F. Lummis described the Petrified Forest as "an enchanted spot...to stand on the glass of a gigantic kaleidoscope, over whose sparkling surface the sun breaks in infinite rainbows."

NAMED AFTER *GAME OF THRONES*
This HBO series has become so popular that scientists are naming new species after its characters. A sea slug has been named *Tritonia khaleesi* in honor of the Khaleesi (Daenerys Targaryen), and two species of ants are *Pheidole viserion* and *Pheidole drogon* after two of her dragons, Viserion and Drogon.

MORE SCIENCE BEHIND TOYS

PLAY-DOH

Although Hasbro won't reveal its exact recipe, Play-Doh is essentially wheat starch and warm water with some lubricants and preservatives. The water makes the wheat starch swell and become supple, giving the Play-Doh its moldable quality. (That's why it gets hard and crumbly if left out: The water evaporates, and the wheat starch shrinks and loses its flexibility.)

SUPER BALL

Polybutadiene is a synthetic rubber that's brittle when cold and goo when hot, neither of which is useful in a ball. But add sulfur, cook the rubber at 330°F while adding 3,500 pounds of pressure per square inch, and it becomes super-bouncy. Size matters: The Super Ball must be molded to about two inches in diameter, or the heat and pressure process won't work.

LEGO BRICKS

Giant hoses suck different-colored plastic granules from trucks into three-story-high metal silos. They're fed into molding machines, heated to 450°F, then fed into hollow LEGO brick molds. The machine applies hundreds of tons of pressure to make sure each brick has the perfect shape. Then the bricks are cooled and ejected. But you have to put them together.

STRANGE SCIENCE

DNAliens

Ever notice how most of the aliens on *Star Trek* are humanoids? The writers eventually came up with a reason for this (other than budgetary constraints) in a 1993 episode of *The Next Generation* called "The Chase" about an ancient alien race that seeded hundreds of worlds with humanoid DNA. And the proof is hidden inside our own DNA. Pure sci-fi, right?

In 2013 two researchers from Kazakhstan, Vladimir I. shCherbak and Maxim A. Makukov, reported that they discovered something unusual while studying DNA: a "mathematical and semantic message" deeply embedded in our genes. Could it have been placed there by aliens who seeded our world? "This code," the researchers wrote, "is the most durable construct known. Therefore it represents an exceptionally reliable storage for an intelligent signature." They call their finding a "biological SETI" (Search for Extraterrestrial Intelligence) and maintain that once our genome is fully charted, we may discover that the proof that there is intelligent life elsewhere may have been hidden inside us this whole time.

HANGOVER "REMEDIES"

You can't cure a hangover once you've got one—it's that simple. These traditional remedies don't work, but some are so disgusting that at least they'll take your mind off of being hungover.

- Swallow six raw owl eggs in quick succession.

- "Hangover Breakfast"—black coffee, two raw eggs, tomato juice, and an aspirin.

- Jackrabbit tea: Take some jackrabbit droppings; add hot water to make strong tea. Strain the tea, then drink. Repeat every 30 minutes until the headache goes away or you run out of droppings.

- Whip yourself until you bleed profusely. The loss of blood won't cure the hangover, but it will (1) make you groggy, and (2) serve as a distraction.

- Drink the sugary juice from a can of peaches.

- Add a teaspoon of soot to a glass of warm milk (hardwood soot is best). Drink.

- Spike some Pepto-Bismol with Coca-Cola syrup from the drugstore, or with a can of day-old Coke, and down the hatch it goes.

SAFECRACKING SCIENCE

A safe really can be opened using an ordinary stethoscope, but it's much more tedious than it's usually depicted in movies. Modern safes are much quieter than older models, so stethoscopes have given way to electronic listening devices.

So what are safecrackers listening for? If you thought they were trying to hear the tumblers tumbling, think again.

- There's a piece of hardware in the wheel pack called a drive cam. It, like the wheels in the wheel pack, has a notch in it.

- By turning the dial on the safe, the safecracker can find the location of this notch by listening for two clicks. The first click indicates where the notch begins, and the second indicates where it ends. Let's say the dial is numbered from 0 to 99: The two clicks might be heard at number 15 and 25 on the dial.

- When the dial is turned a certain way, the spacing between the two clicks will shrink ever so slightly, say from 15 and 25 on the dial to 18 and 22. But— and this is important—the space between the clicks

shrinks only when you begin the procedure from certain numbers on the dial. The trick is finding out which numbers, because each one is a number in the combination.

- The only way to find all the numbers in the combination is by repeating the procedure over and over again, using every third number on the dial as a starting point. If the dial is numbered from 0 to 99, for example, you start the procedure at 0, then 3, 6, and so on, until you reach 99 on the dial (that's 33 times in all).

- This trick doesn't reveal the order of the numbers in the combination, but in a three-number combination there are only six possibilities. Once the numbers are revealed, opening the safe is easy.

 5 Elements That Are Liquid
at Room Temperature

1. Mercury
2. Caesium
3. Francium

4. Gallium
5. Bromine

Canada's Oddest Museum?

Have you ever dreamed of seeing a gopher dressed as a Mountie? Or maybe you've fantasized about gophers working as hairdressers and styling each other's locks. How about a gopher dressed as a preacher or an angel gopher with a halo and harp floating in the air above his head? At the Gopher Hole Museum in Torrington, Alberta (just north of Calgary), these visions are all on display for you to behold and admire. About the size of a garage, the museum displays 44 dioramas with 71 stuffed gophers (really Richardson's ground squirrels, to be accurate) elaborately dressed as townspeople doing a wide range of activities, including bank officer, robber, and firefighter. Don't miss: The gopher smith hammering at his anvil; the clown gopher clutching his balloons; a '50s-style female gopher showing off her poodle skirt while holding hands with a young male gopher in a leather jacket.

Green City:
REYKJAVIK

POPULATION: 130,000

HOW GREEN IS IT? In the 1970s Iceland relied on imported coal for 75 percent of its energy. Today all of its electricity is produced from hydroelectric and geothermal power. The hydropower source is flowing water from melting ice that turns turbines to make electricity. The geothermal power uses the heat and steam of Iceland's volcanoes to do the same. The only fossil fuel the city uses is for its cars and fishing fleets.

But Icelanders even consider that to be too much: To get down to zero use of fossil fuels, Reykjavik is working on a changeover to cars and ships fueled mainly by electricity and hydrogen. In 2003 a hydrogen filling station opened in Reykjavik to service hydrogen-powered public buses. By the mid-21st century, Iceland plans to have most of its fishing fleet running on hydrogen and all of its cars and buses powered by alternative fuels.

* * * * * * *

We're gonna need a bigger spoon:
It's estimated that a teaspoonful of a neutron star
would weigh upwards of six billion tons.

A "Scientific" Documentary

In the feature-length film *Ingagi* (1930), an intrepid British explorer in the jungles of the Congo comes across a tribe of people who sacrifice their "naked ape-women" to gorillas. And the "naked ape-women" have sex with said gorillas. The film even showed a child with tufts of curly hair taped to his body who was supposed to be the progeny of such an encounter. *Ingagi* was marketed as a straight documentary by RKO Radio Pictures, one of the most respected film companies of the day, and it smashed several box office records for a few months before it was revealed as a fraud and theaters stopped showing it. One scene showed the discovery an entirely new species of animal called a "tortadillo." It was later revealed to be a turtle with wings, scales, and a tail glued to it.

The film was a precursor to RKO's megasmash hit of 1933—*King Kong*. And it was one of the first documentaries with sound—which was just becoming common in films—in the form of the now ubiquitous documentary narrator.

RUBIK'S CUBE ROBOT

The Rubik's Cube became an international phenomenon in the 1980s, but it hasn't entirely faded. "Speedcubing" became so popular that the World Cube Association was founded in 2003 in order to stage competitions and document speed records. But our puny human hands can't come close to competing with robots that can solve a Rubik's Cube in less than a second.

- The first of these robots is an unnamed creation that requires a specially built Rubik's Cube for scanning and spinning purposes. First, the cube is scanned on all six sides via camera. Then the custom cube is put into a glass-and-steel contraption that spins the cube into its finished stage. While the spinning itself takes less than a second, the scanning takes much longer.

- Designer Mike Dobson created a robot that analyzes *and* solves the cube in about three seconds. The CubeStormer 3, made almost entirely out of Legos, uses a smartphone to analyze and solve the cube at the same time.

THE GRIFTERS

In 1582 scientist John Dee (see page 129) started a partnership with Edward Kelley, a medium who claimed he could contact angels and spirits by gazing into a crystal ball. Kelley was also a failed lawyer who had been convicted for fraud, forgery, and counterfeiting. Kelley conned Dee into thinking he could help him in his quest to talk to the spirit world, and an unusual partnership was born.

Together, Kelley and Dee became the Siegfried and Roy of their day. Dee supplied the magical knowledge, while Kelley played the charismatic front man. They claimed they were able to contact angels, especially one divine being called Uriel, and they dubbed their system Enochian Magick, after the biblical character Enoch. They toured as far afield as Poland and Bohemia, performing for princes and kings. It was only when Kelley told Dee that Uriel recommended that they indulge in wife swapping that Dee recognized his colleague as a con artist. Their partnership ended in 1589, by which time Kelley had made a fortune and earned a knighthood. Dee, on the other hand, hadn't made a penny.

Dee's final years were spent in near-poverty and obscurity until he died in 1608. His contemporary William Shakespeare modeled the philosopher-magician character of Prospero in *The Tempest* on John Dee.

DUMB PREDICTIONS

"This 'telephone' has too many shortcomings to be seriously considered as a means of communication. The device is inherently of no value to us."

—WESTERN UNION INTERNAL MEMO, 1876
(after Alexander Graham Bell offered to sell them the rights to the telephone)

MORE FRANKENFOODS

ARCTIC APPLES won't turn brown. The apples, which are produced by Okanagan Specialty Fruits, have been bioengineered so that the enzyme that causes the browning is turned off. They're currently being used by many fast-food companies, and are expected to hit retail stores in the near future.

ATLANTIC SALMON grow faster and larger than ever before. According to the watchdog group GreenAmerica, the fish are "engineered with a growth-hormone-regulating gene from a Pacific Chinook salmon and a growth promoter from an ocean pout [an eel-like fish] to make it grow to a larger size at a faster rate."

CLONED BEEF. Has any of the milk or beef you've consumed come from the offspring of a cloned animal? According to Food Standards Australia and New Zealand (FSANZ), "Food products from [cloned animals'] offspring are almost certainly in the food supply." Are they safe? According to FSANZ, "The U.S. Food and Drug Administration, European Food Safety Authority, and Japan Food Safety Commission...have concluded that food products from cloned animals and their offspring are as safe as food products from conventionally bred animals."

PESTICIDE-RESISTANT WHEAT, developed by Monsanto and Bayer, also has GreenAmerica concerned because "GMOs focused on pest- and weed-resistance have started to fail, as the pests are adapting to GMOs and related chemicals, evolving into superbugs and superweeds."

RAINBOW PAPAYAS. After the ringspot virus severely damaged Hawaii's papaya crop in the 1990s, a plant pathologist named Dennis Gonsalves inserted the ringspot's genetic material into the papayas to create a ringspot-resistant fruit. Good news: it worked. Bad news: the papaya pollen drifted away and contaminated all the other papayas, so if you want to buy an organic papaya grown in Hawaii, good luck.

"A potato can cross with a different strain of potato but, in 10 million years of evolution, it has never crossed with a chicken. Genetic engineering shatters these natural species boundaries, with completely unpredictable results."
—Michael Khoo (from a letter published in the *Toronto Globe and Mail*)

Mixed-Up Heritage

In 2006 Henry Louis Gates Jr., director of Harvard University's W. E. B. Du Bois Institute for African and African American Studies, produced and hosted a PBS documentary entitled *African American Lives*. Gates asked several prominent African Americans to submit to genealogical DNA testing to help trace their family trees. They all had interesting results.

OPRAH WINFREY believed she was of Zulu descent (the Zulu people now reside in South Africa). She also believed she had no Native American or European ancestors. She was mostly wrong. The tests showed her to have 89% sub-Saharan African, 8% Native American, 3% East Asian, and 0% European heritage. Regarding her sub-Saharan ancestors, Gates told her that she descends from the Kpella tribe in what is now the West African nation of Liberia; the Bamileke people, in modern-day Cameroon; and the Nkoya people in Zambia.

Comedian **CHRIS TUCKER** guessed he was descended from a tribe in modern-day Ghana. Wrong. His test showed 83% sub-Saharan African, 10% Native American, and 7% European descent. And the African link was to the Mbundu tribe in present-day Angola. Gates said the link was so strong that it was likely that a direct ancestor of Tucker's had been taken into slavery in Angola sometime in the 1700s. (The show featured Tucker going to Angola and meeting with his distant relatives.)

Music mogul **QUINCY JONES** said that family stories told of Native American ancestry, and thought he had probably a little European blood, too. Result: 66% sub-Saharan African, 34% European, and 0% Native American. Jones was shocked to learn that his father's line showed *only* European descent, while his mother's showed a connection to the Tikar people of modern Cameroon.

The word *evolution* never appears in the first edition of *On the Origin of Species*. Charles Darwin didn't use the term until he revised the text for its sixth printing, in 1872.

IT'S PRIMAL

Bottling up your emotions isn't good for you; that's obvious. But according to advocates of primal therapy, the effects can go far beyond headaches and grinding teeth. Repression, they claim, can cause a slew of physical and mental maladies. Primal therapy was pioneered by Arthur Janov in the early 1970s. Janov taught that we are all marked by pain felt early in life—usually a lack of love in childhood—and that internalized pain manifests itself in a range of illnesses, including high blood pressure, cardiac arrhythmia, ulcers, phobias, depression, and autoimmune disorders like allergies and asthma. By regressing to an infantile state, the patient can confront and release this pain, exorcising it with a cathartic primal scream.

The therapy has been controversial. Its claims of a single universal root cause for a vast range of seemingly unrelated illnesses (and of the efficacy of a single treatment against all of them) strikes many diagnosticians as overly simplistic. But the method has some serious cultural clout, thanks to devotees like John Lennon, who wrote some of his most powerful solo material while undergoing treatment. "Shout," the 1984 hit by Tears for Fears, is about primal therapy. So shout! Shout! Let it all out!

HOW SOAP IS MADE

Oil and water don't mix; they repel each other like opposite ends of a magnet. When you wash your skin with water alone, the oil (sebum) in your skin repels the water and keeps it from cleaning the skin effectively. That's where soap comes in.

Primitively speaking, soap is oil plus alkali. For centuries, that meant fat plus lye. American colonists and pioneers saved fat scraps from cooking. They also saved the ashes from their fireplaces, which they placed in a barrel with a spigot at the bottom. Water poured over the ashes and left to soak would form lye, which was then drained off from the bottom. The cooking fat was rendered in a vat over a fire, then the lye was added. After much stirring and cooking, a chemical reaction would take place, and soap was the result. Too much lye, and the soap would be harsh on the skin. Too much fat, and the soap is greasy. The newly formed soap would then be poured into boxes to harden and cure for several months.

For more soap stories, go to page 381.

WHAT'S COOKING?

In the late 1800s, Dr. Frank Buckland caused quite a stir in England. He predicted a time would come when Britain would be overpopulated. The country's farms would not be able to feed them all, so he set up a society to find tasty new foods. On the menu: silkworms, beavers, parrots, and other unusual items.

Buckland got his weird tastes from his father, who spent his life studying animals. The senior Buckland claimed to have eaten through the entire animal kingdom, all in the name of science. Frank's dad invited some of the leading scientists of his day to try out new recipes. Mice on buttered toast was a hit. Hedgehog was "good and tender," but crocodile was a disaster; none of his guests could gulp it down.

As an adult, Frank became a doctor. He often treated sick animals at the London Zoo. Sometimes he could not save his patients, so...he ate them. That's how Elephant Trunk Soup, Panther Chops, and Rhinoceros Pie ended up on his dinner table. After a fire at the zoo, Dr. Buckland served "Accidentally Roasted Giraffe" to his guests.

On July 12, 1862, Buckland's society held its first official dinner. The menu included Bird's Nest Soup, Sea Slug Soup, and Deer Sinew Soup. Buckland thought the soups all tasted like glue. The Kangaroo Stew was "not bad, but a little gone off."

Buckland's exotic meals didn't catch on in England. Tibetan yak steaks and Japanese sea slugs weren't appetizing to most people. But his society did bring ostrich, water buffalo, and bison farming to Britain. Among the favorite dishes in Britain today: Toad in the Hole—made (thankfully) with sausages, not toads.

 7 Steps of the Scientific Method

1. Observation
2. Statement of a problem or question
3. Formulation of a hypothesis, or a possible answer to the problem or question
4. Testing of the hypothesis with an experiment
5. Analysis of the experiment's results
6. Interpretation of the data and formulation of a conclusion
7. Publication of the findings

PATENTLY WEIRD VEHICLE PATENTS

INVENTION: Collapsible Riding Companion (patent no. 5,035,072)

INVENTOR: Rayma E. Rich; Las Vegas, NV

DETAILS: Afraid to drive alone at night? Want to drive in the carpool lane? Just want a little company? The Collapsible Riding Companion is a dummy head and torso complete with a full head of hair, T-shirt, and zippered jacket that rides "shotgun" wherever you need to go. The head and torso collapse into a lightweight rectangular case for easy storage.

INVENTION: System for Protecting Against Assaults and/or Intrusions (patent no. 4,281,017)

INVENTORS: Yari Tanami, Yoav Madar; Gedera, Israel

DETAILS: If you're worried that your passenger might assault you, then this is the invention for you. Electrodes over the front and back passenger seats are connected to a high-voltage ignition coil. If the driver is threatened, a foot switch sends a charge of electricity through the offending passenger strong enough to temporarily stun them. Leaving your car parked in a bad neighborhood? Set the system on "FRY" intruder detection mode and give some unsuspecting burglar the shock of his life.

FIVE FREAKY FACTS ABOUT...
OUTER SPACE

- If you could capture a comet's entire 10,000-mile vapor trail in a container, the condensed vapor would occupy less than one cubic inch of space.

- In any given year, about 26,000 meteorites land on Earth's surface, the vast majority dropping into the oceans.

- The most common color of star in the universe is not white but red.

- Earth travels through space at 66,600 miles per hour—eight times faster than the speed of a bullet.

- Oh no! The Sun is a middle-aged star that has only about 5 million years before it dies.

PSYCHED FOR CYCADS

They look like a cross between a palm and a fern, with a stout trunk and a crown of featherlike leaves across the top. But these ancient plants—called *cycads* (pronounced "SY-kads")—are more closely related to gingko trees and conifers.

- Cycads have been growing on this planet for more than 300 million years, making them among the oldest species of any kind still living in the world.

- Cycads contain BMAA, a paralyzing neurotoxin. But native peoples in Australia, Africa, and North America found ways to leach out the poison and turn the starchy stems into edible flour.

- The Seminole Indians of Florida called cycads the "white bread plant." Their entire diet was based around *sofkee,* a pudding made from its starch. When Confederate soldiers garrisoned in Florida during the Civil War ran out of provisions, they tried to create their own version of sofkee. Unfortunately, they skipped the soaking process that removed the plant's poison—and hundreds of soldiers died.

- White settlers in Florida eventually learned the Seminole process and made a cooking powder they called arrowroot starch, or *coontie*, which was what the Florida cycad was called. During World War I, coontie mixed with beef broth was the only food that soldiers who'd been gassed could stomach.

- The Japanese word for cycad is *sotetsu*. Cycad nuts were eaten as a food of last resort during famines, and a particularly bad famine in the 1920s is still referred to as *sotetsu jinkoku*, or "cycad hell."

- The Japanese sago "palm" is perhaps the best-known cycad in the world (though misnamed—it isn't a palm).

- A great petrified forest of cycads used to lie just outside Minnekahta, in the Black Hills of South Dakota. It was once a national monument until fossil hunters stripped away all of the visible specimens and sold them to museums and collectors.

- Cycad seeds look like pinecones, and can weigh as much as 90 pounds.

- The largest cycad alive today is a Hope's cycad located in Daintree, Australia. It's 1,000 years old and 65 feet high.

THE SHOCK THERAPY EXPERIMENT

In the 1970s and 1980s, South African army psychiatrists performed experiments on members of the nation's all-white military who were—or were suspected of being—gay or lesbian. The experiments involved encouraging subjects to fantasize about someone of the same sex and then delivering powerful electric shocks to them. If repeated treatments didn't "cure" them, other methods were tried, including hormone therapy and chemical castration. The exact number of men and women subjected to these experiments is unknown—some estimates put it as high as 900. News of the experiments wasn't made public until the end of South Africa's apartheid era in 1994.

UPDATE: One psychiatrist believed to be involved in these experiments, Dr. Aubrey Levin—known as "Dr. Shock"—was allowed to emigrate to Canada in 1995, where he became a professor of psychiatry at the University of Calgary. Levin's psychiatry license was revoked in 2010, after he was charged with sexually assaulting male patients; he was found guilty and ordered to serve a five-year prison term.

Let There Be Light

In 1879 passengers on trains traveling from New York
to Philadelphia were in for an incredible sight. On those
December nights, all of the towns the trains passed were
bathed in darkness...except one. Menlo Park, New Jersey—
home to Thomas Edison and his "invention factory"—
sparkled with light. It was all part of Edison's plan to draw
attention to himself and his inventions. "The Wizard of
Menlo Park," as the press had dubbed him, worked on the
stunt for months. He bragged that he intended to light
whole cities with his electrical system and that it was only
a matter of time until gaslight, which he called dirty and
unsafe, became obsolete.

Edison laid eight miles of underground wire across
half a square mile of his Menlo Park laboratory property.
His workmen planted rows of white, wired posts to hold
the thousands of lightbulbs his factory had mass-produced.
Glass globes covered the bulbs. The old library annex
was converted into a central power station containing 11
generators. When the trains passed by, Edison turned a
wheel in the power station and flooded the barren fields

with a brilliant array of twinkling streetlights. In addition to the impressive outdoor display, hundreds of lamps installed in his New Jersey research facility sprang to life. It was the most incredible display of artificial light that the world had ever seen.

That winter, Edison was the toast of two cities. Congressmen, dignitaries, bankers, stockbrokers, and celebrities traveled to see his "Fairy-land of Lights." The publicity stunt worked. He got the go-ahead to bring electricity to Manhattan.

Less than three years later, in 1882, Edison had installed a central generating station that was humming away on New York's Pearl Street. And as one last treat for the city's residents, Edward H. Johnson, a longtime associate of Edison's, put a Christmas tree in the window of his New York City home and decorated it with 80 red, white, and blue lights. The electric age had begun.

IT'S THE BLOOMIN' ALGAE

A fascinating spectacle sometimes seen in oceans and rivers is a red tide, or more accurately an "algal bloom." The water is tinted a brilliant red color by the overgrowth of an algae species that is normally microscopic. Algal blooms form when too many nutrients like nitrogen and phosphorus collect in water, causing the algae to reproduce so quickly that the ecosystem is overtaken. When these blooms coat the water's surface, they block sunlight, hog the oxygen, and kill off underwater plants and animals. Some red algal blooms also produce toxins that harm aquatic life and humans. Mammals can become ill or die if they swim in the water or eat shellfish harvested there. At night, red tidewater turns an amazing bioluminescent blue. Why? When waves or boats disturb the microscopic organisms, it causes a chemical reaction that creates a flash of blue light, which is multiplied by billions of cells within the red tide. Algal blooms may hang around for months, wreaking havoc on the environment and the area's tourism trade.

STRANGE MEDICAL CONDITION

SUBJECT: Shanya Isom of Memphis, Tennessee

CONDITION: Unknown condition affecting hair follicles

STORY: In September 2009, Isom, a 28-year-old student at the University of Memphis, had an allergic reaction to the steroids she was given to treat an asthma attack. Initially, the reaction caused her skin to itch all over her body...but then it got much worse: she started growing oddly shaped, stiff, sharp black growths all over her body. Two years later, after seeing dozens of doctors, all to no avail, she sought treatment at Johns Hopkins Medical Center in Baltimore. There, doctors finally figured out what was happening. Isom's hair follicles were producing 12 times the number of cells they normally do, and they weren't producing hair cells—they were making nail cells, like the ones produced for fingernails and toenails. She was basically growing sharp black fingernails all over her body. The doctors informed Isom that she was the only person ever known to have such a condition...and they had no idea what was causing it. They were able to alleviate her symptoms to a degree, but unfortunately, she continues to suffer from the condition today.

DUNE TUNES

In the 13th century, while traveling through the Gobi Desert, explorer Marco Polo heard eerie sounds coming from the sand dunes around him. He described the noise as "all kinds of musical instruments, and also of drums and the clash of arms." After hearing the mysterious noises, Polo came to the "logical" conclusion that he must be in the presence of evil spirits. These days, we know that all that music was nature, not spirits. Of all the sand dunes in the world, only a few have the ability to "sing" in the ways that so startled Marco Polo. Beach sand sometimes makes brief squeaking noises, but it's rare to find dunes that produce the magnificent instrumentals Polo described. There are actually only about 30 singing sand dunes on earth, including:

- The **Kelso Dunes**. Located in the Devil's Playground area of the Mojave National Preserve, the **Kelso Dune**s rise as high as 650 feet.
- The sounds of the **Dumont Dunes**, also located in the preserve, were filmed for the PBS *Nova* episode "Booming Sands." According to researchers, the **Dumont Dune**s "sing" the note of G.
- The **Eureka Dunes** in Death Valley National Park. Rising 680 feet from the valley floor, the dunes are also thought to be the tallest in North America.

  Go to page 354 to learn more about these musical dunes.

WHEN A BLACK HOLE THROWS UP

Black holes suck, but they also blow. For a long time, astronomers have known what they suck: matter and light from nearby stars. But the blowing has been something of a mystery. It's known that black holes create intense heat that is emitted as X-rays. These "space jets" can shoot out at more than 400 million miles per hour, or about two-thirds the speed of light.

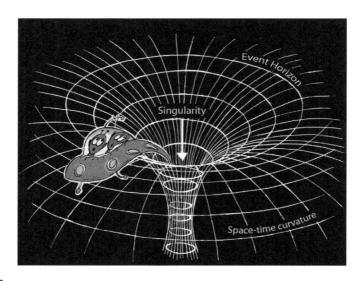

But for some unknown reason, the jets are positively charged. To find out why, astronomers used the *XMM-Newton*—an X-ray space observatory launched by the European Space Agency in 1999—to get a closer look at 4U1630-47, a relatively small black hole not too much larger than our Sun. "We've known for a long time that jets contain electrons," wrote Australian astronomer James Miller-Jones, "but they haven't got an overall negative charge, so there must be something positively charged in them, too." When the astronomers examined the X-rays, they expected the culprit to be something exotic like antimatter, but it turned out to be two elements that are abundant here on Earth: nickel and iron. For some reason, the black hole rejected them. The reason why is still a mystery.

> "Science and technology revolutionize our lives, but memory, tradition and myth frame our response."
> —**Arthur Schlesinger (historian)**

Animals with Heart

Like humans, most other creatures need hearts to stay alive—the exceptions are jellyfish and coral, which don't have hearts. But that doesn't mean that all hearts work like ours do.

- Instead of circulating oxygen-rich blood back to the lungs, a fish's heart sends the blood to its gills.

- Frogs have three-chambered hearts: two atria and just one ventricle.

- Insects have hearts (of sorts), but their circulatory system is open. This means that bug blood (called hemolymph) flows throughout the insect's body— not in arteries and capillaries like ours—propelled by a vessel in its abdomen that functions as a heart.

- The blue whale is the largest creature on earth, so it stands to reason that the whale's heart is the largest, too. A blue whale's heart can weigh 1,300 pounds... about half the size of the average compact car.

- A hovering hummingbird's heart flutters at 1,200 beats a minute or more.

- The giraffe has the highest blood pressure in the animal world because its heart must pump blood with extra force to overcome the gravity associated with a head so high. That doesn't mean the giraffe

is a big-hearted fellow, though—its heart is as big as any animal its size. Instead, the walls of a giraffe's heart are thicker and more powerful. And to accommodate all that extra pressure, the giraffe's blood vessels also thicken as the animal (and its neck) grows.

• During hibernation, a bear's heart rate drops to about 22 percent of the number of beats it needs when it's active—about 19 beats per minute in hibernation versus the normal 84 beats per minute while it's doing things like raiding your campsite looking for marshmallows.

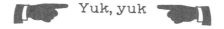 Yuk, yuk

Q: Did you read that new book about antigravity?

A: Yeah, I couldn't put it down!

MORE TREK*NOLOGY

STARBASE

Trek*nology: When the *Enterprise* needed repairs, or the crew needed some R&R ("rest and relaxation"), they set course for the nearest starbase—a floating space city that supported hundreds, sometimes thousands, of people with food, housing, and entertainment.

Technology: On May 14, 1973, the United States launched the space station *Skylab*, the home base in space for U.S. astronauts until February 1974. The Russian space station *Mir* circled Earth from 1986 to 2001. In 1998 the International Space Station began its service as Earth's "starbase."

SHUTTLECRAFT

Trek*nology: When Kirk and his crew needed to move people and equipment from the *Enterprise* to a planet's surface, they often used the shuttlecraft—a small space vehicle that could go from ship to planet or starbase, and back again.

Technology: Fifteen years after the first *Star Trek* episode, NASA launched the first space shuttle. There were 135 space shuttle missions from 1981 through 2011, including 37 dockings at the International Space Station.

"Scientific" Theory: CALIFORNIA ISLAND

In the 1620s, respected British mapmaker Henry Briggs published the most detailed map of North America to date. It became one of the most respected New World maps in Europe and helped to promote a theory that prevailed for more than a century: that California was a long, narrow island off the west coast of North America. (The error stemmed from Briggs trusting an earlier and equally wrong Spanish map.) And even though it was soon contested by explorers to the region—and by the early 1700s was proven to be wrong by people who actually went overland to California—the "California is an island" theory persisted well into the 1700s. It was finally dealt a death blow in 1747 when King Ferdinand VI of Spain formally decreed that California was not an island.

A Viking Surprise

Trying to figure out what life was like more than 1,000 years ago is extremely difficult...but science can make it easier. Take the case of 14 Viking skeletons that were discovered in England with items such as knives, swords, and shields in their graves. A casual observer might assume that these warriors were all men. Not so fast, says Shane McLeod, an Australian archaeologist who tested the skeletons' DNA. His goal: to add light to a recent theory that the invading Vikings, who arrived circa A.D. 900, weren't necessarily hordes of barbaric men, but rather married couples who (violently) colonized western Europe. So what did McLeod's DNA tests reveal? Of the 14 Vikings, six turned out to be female, seven were male, and one was indeterminate. That finding, along with some

recent discoveries of Viking-era Norse jewelry found in England, has led McLeod to "caution against assuming that the great majority of Norse migrants were male." Of course, more research needs to be done on larger sample groups, but if he's correct, a third to a half of the Vikings who invaded England were lady Vikings.

THE NUCLEAR BOY SCOUT

In 1994 David Hahn, a 17-year-old from Detroit, Michigan, learned that it was possible to find small amounts of radioactive material in common store-bought items. *Americium-241*, for example, is found in most smoke detectors. The science-minded Boy Scout set his sights on earning the (now-discontinued) "Atomic Energy Badge"...by building a nuclear reactor in his mother's backyard shed.

First, he had to create a "neutron gun" (the nuclear reactions that power most nuclear plants are set off by bombarding a radioactive element, usually uranium, with neutrons). Hahn collected americium-241 from hundreds of smoke detectors and packed them into a hollow piece of lead with a tiny opening. Radiation can't pass through lead, so the radiation from the americium-241 could now only escape through that pinhole—as a focused beam. Hahn covered the hole with a thin strip of aluminum, which reacts to radiation by ejecting neutrons. He now had his "neutron gun."

Over the next several months, Hahn attempted to create a nuclear reaction by shooting different radioactive substances with his neutron gun. That included thorium-232, which can be found in gas camping lanterns, and beryllium, which he stole from a chemistry lab. After that, he acquired some pitchblende, a type of rock that contains small amounts of uranium. Hahn never succeeded in creating a nuclear reaction...but he did create extremely dangerous levels of radiation—more than 1,000 times normal levels. The fiasco finally ended when police stopped Hahn one night in August 1994, and he told them he had radioactive substances in his car. The FBI and the Nuclear Regulatory Commission were immediately called, a Federal Radiological Emergency Response Plan was initiated, and Hahn's mother's property was designated a hazardous materials site. The shed, along with all of its contents, was buried at a radioactive waste disposal site in Utah. Hahn refused medical evaluation, despite having been told that he had been exposed to more radiation than a person can safely endure...in an entire lifetime.

Postscript: In August 2007, the 30-year-old Hahn was arrested. He pled guilty...to stealing several smoke detectors. He was sentenced to 90 days in jail. Then, in 2016, he died. Cause of death: unknown.

WHISKER SCIENCE, Part II

- The width of a cat's outstretched whiskers is usually the same as the width of the body. So cats use their whiskers to measure the diameter of holes or other openings to make sure they're wide enough to enter without being trapped. When a cat overeats and gains too much weight, though, his whiskers stay the same size. So a fat cat may misjudge the size of its body and get stuck in a hole—one good reason not to overdo the treats.

- Cats also have whiskers on the backs of their two front paws. These are shorter than the ones on a cat's cheeks and help it walk over uneven ground without stumbling and to determine the size and position of captured prey.

- A cat's whiskers should never be trimmed because without its whiskers, a cat can get disoriented in the dark.

- The feline fetus develops whiskers before any other hairs. And when kittens are born, they're blind and deaf, but the touch sensors on their whiskers are fully operational.

WHERE DID OUR MOON COME FROM?

Not long after humans finally walked on the Moon and samples were brought back and studied, astronomers W. K. Hartmann and D. R. Davis proposed a new theory, called the giant impact theory. When Earth was just 50 million years old—some 4.5 billion years ago—a planetary object the size of Mars slammed into it (a slightly off-center hit, they said). The enormous impact resulted in a huge amount of debris—both from the object and from Earth's mantle—being thrown into space. Much of it was affected by Earth's gravity and fell into orbit. This material eventually coalesced into the Moon. This would explain not only the different makeup of the Moon but also its lack of a large iron core, since the Earth's mantle was by this time largely free of iron (because the iron was in the core). Today, the giant impact theory is the most widely accepted.

THE BIRTH OF E-MAIL ▸

Ray Tomlinson was just goofing off at work one day when he created the way a lot of us communicate these days. After graduating from MIT in 1965, he went to work at Bolt, Beranek and Newman, the company that had contracted with the U.S. government to build ARPANET— the experimental military communications system that would later become the Internet.

In 1971 Tomlinson was trying to figure out a way to send messages to other engineers on the project. He knew of a program that could send messages between users of the same ARPANET machine, but he also knew of another program that could send files from one remote computer to another—so why not messages? He tinkered some more and figured out how to use them both to get what he wanted. And he chose a symbol for message address lines to denote mail sent through ARPANET to remote machines: today's ubiquitous @.

Electronic mail caught on like wildfire. It wasn't the programming that was the breakthrough, it was the idea. Suddenly two users could send terse, information-filled messages to each another without the need for social niceties or for both to be available to chat at the same time. The number of e-mail messages sent grew by leaps and bounds on ARPANET. By 1973 a study found that 75 percent of all traffic on ARPANET was e-mail. Today we send millions of e-mails every day—it's changed the way we do business, talk to one another, and even the way we think.

Tomlinson didn't consider these messages to be a big deal; in fact, as *Forbes* reported in 1998, a BBN coworker said that when Tomlinson showed him his work, he said, "Don't tell anyone! This isn't what we're supposed to be working on."

DUMB PREDICTIONS

"Two years from
now, spam will
be solved."

—MICROSOFT FOUNDER
BILL GATES, 2004

SUCCESS
in
DE FEET

A key to athletic success is hooking up the right athlete to the right sport, and even small structural differences in feet can determine whether someone can be a star at the 100-meter hurdles or a powerhouse on the tennis court. Talent scouts searching for speedy quarterbacks, sprinters, or base-stealing ballplayers might do well to examine a candidate's big toe. For most of us, the big toe isn't as long as the next one, but some people have big toes that protrude out beyond the second toe. These fortunate few have an advantage over the rest of us when they need speed. They can lean their weight onto their big toe to push off and get a fast start. The second toe is not as strong and can only exert about half as much force.

Others have a unique advantage in the first metatarsal bone, which is attached to their big toe. If the first metatarsal bone hangs lower than the other metatarsals

(the bones to the other toes), then the big toe will also hang lower than their other toes. Athletes with a low first metatarsal can also put weight on their big toe, pushing off for a fast start.

If your feet tend to roll outward and make your arch more stiff and rigid, you might want to try out for track or volleyball. Rigid feet are good levers that make running and jumping easier. If your feet tend to roll inward and your arches are extremely flexible, that could give you an advantage at tennis or aerobic dancing. Flexible feet are better at handling constant changes in direction with quick, short pivots.

People with flat feet and lower arches have their own advantages. They usually fall in the category of flexible feet with good range of motion. Even just plain big feet can be an advantage in swimming (think flippers).

KINETIC SCULPTURE TRIATHLON

The only race in the United States (we think) whose participants are allowed help from interplanetary beings takes place every May in Humboldt County, California. The Kinetic Grand Championship requires that teams create moving sculptures and pedal them 42 miles from Arcata to Ferndale through water, mud, and sand. The #1 rule: It's mandatory to have great fun. Rule #2: Sculptures must be powered by humans, but "it is legal to get assistance from the natural power of water, wind, sun, gravity, and friendly extraterrestrials (if introduced to the judges)." The contest has been going on since 1969, when sculptors Hobart Brown and Jack Mays raced their creations (one was a five-wheeled cycle) down Main Street in Ferndale. Today, each sculpture has a pilot to steer, a pit crew for maintenance, and "peons" who handle the rest, including any necessary bribing of spectators and judges. Past entries include a giant lobster, a dragon boat, and a Day of the Dead decorated taco truck. This contest doesn't just recognize the team that is the fastest. Sculptures can win for taking the middle spot (the Mediocre Award), for being the first to break down (the Golden Dinosaur), and for not cheating (Ace the Race). However, the organizers make it clear that the race condones cheating only "if it is done with originality and panache."

IT'S SCIENCE!

"All truths are easy to understand once they are discovered; the point is to discover them."
— *Galileo Galilei*

"Science is simply common sense at its best—rigidly accurate in observation and merciless to fallacy in logic."
— *Thomas Huxley*

"There are many hypotheses in science that are wrong. That's perfectly all right; they're the aperture to finding out what's right."
— *Carl Sagan*

"The good thing about science is that it's true whether or not you believe in it."
— *Neil deGrasse Tyson*

"The greatest discoveries of science have always been those that forced us to rethink our beliefs about the universe and our place in it."
— *Robert L. Park*

"Every great advance in science has issued from a new audacity of imagination."
— *John Dewey*

ANOTHER VIRGIN BIRTH

WHO? Sungai, a Komodo dragon

WHERE? The London Zoo, England

WHAT? Born and raised in captivity, Sungai had not interacted with a male Komodo dragon in more than two years. Even so, she laid a clutch of 22 fertilized eggs in 2005. Four of the eggs hatched, all male, and did not contain DNA that would have come from a daddy dragon. The birth stunned scientists, who had not known Komodo dragons could reproduce asexually. It is unclear whether this occurred through cloning—which, researchers note, should have produced all females, not males—or perhaps a process called *selfing*, in which an animal's body stores some cells that act like sperm and others that behave like eggs.

WHY? Here's what the scientists think: Komodo dragons are the biggest lizards on earth, but in the wild, they are confined to one small part of the world—a few volcanic islands in Indonesia. They are also endangered. All this makes survival a tricky business. But with this unique reproduction ability, females could produce offspring all on their own—even colonize a new island if necessary—all without the help of a partner and perhaps saving the species from total extinction.

WEIRD ENERGY:
SEWAGE

You are practicing good hygiene when you flush your toilet. You're right to want to dispose of your body's waste products as quickly and efficiently as possible. But your poop might have incredible economic and environmental value. Feces is a key resource in obtaining methane, a natural gas that could be used for heat and energy, similar to natural gas. Park Spark in Cambridge, Massachusetts, along with Norcal Waste in San Francisco, is testing out pilot programs designed to extract as much usable methane as possible from (for now) dog poop. The companies provide dog owners with biodegradable dog waste bags. The doggies provide the waste, the owners fill the bags, and then the energy companies feed the bags into a machine called a "digester," where microorganisms process the dog poop. The byproduct: methane.

ACCIDENTAL DISCOVERY: SAFETY GLASS

In 1903 Édouard Bénédictus, a French chemist, was experimenting in his lab when he dropped an empty glass flask on the floor. It shattered, but remained in the shape of a flask. Benedictus was bewildered. When he examined the flask more closely, he discovered that the inside was coated with a film residue of cellulose nitrate, a chemical he'd been working with earlier. The film had held the glass together. Not long afterward, Benedictus read a newspaper article about a girl who had been badly injured by flying glass in a car accident. He thought back to the glass flask in his lab and realized that coating automobile windshields, as the inside of the flask had been coated, would make them less dangerous. Variations of the safety glass he produced—a layer of plastic sandwiched by two layers of glass—are still used in automobiles today.

SHOWERING ON THE ISS

Astronauts live and work on the International Space Station for weeks or even months on end. The station has a shower, which is pretty fancy considering the astronauts who traveled on space shuttles years before the ISS had to make do with sponge baths and shampoo that didn't need to be rinsed out.

In the absence of gravity, the water used in a shower doesn't fall to the floor. It just floats around inside the stall, which is sealed to prevent the water from escaping into the rest of the space station. Because the water doesn't go down the drain, you don't need as much to take your shower as you would on Earth. You use only about a gallon of water, and instead of moving in and out from under the showerhead, you just grab the floating globs of water and rub them on yourself. When you're finished, there's a vacuum hose attached to one wall that you use to suck up all the drops before leaving the shower. One thing to consider: the shower water is reclaimed water... from the astronauts' sweat and urine.

STRANGE STUDY:
Spiders Get Personal

Most spiders lead solitary lives, but some species, such as *Anelosimus studiosus*, live socially in massive webs. Within these colonies, some spiders are docile and some are aggressive. If the ratio is off, too many docile spiders won't be able to ward off predators, and too many aggressive spiders will cause too much infighting. Either way, the colony collapses. There's a mechanism present to keep these ratios balanced, but exactly what it is has divided the scientific world for decades.

When writing about natural selection, Charles Darwin noted that some individuals may "sacrifice themselves for the common good." This is known as *group selection*, but it hasn't gained much traction because it's never been observed...until now, thanks to a six-year study of *Anelosimus studiosus* conducted by behavioral ecologists Jonathan Pruitt of the University of Pittsburgh and Charles Goodnight of the University of Vermont.

Researchers started three new wild colonies in Tennessee—one with two docile spiders, one with two aggressive spiders, and one mixed. The first year, the docile colony thrived; there was little infighting and the web grew considerably. But it couldn't withstand outside attacks, so by the time the study concluded six years later, the docile colony was gone...and the other two were thriving. How? Group selection. When a colony had a lot of aggressive members but not a lot of external threats, the arachnids would "readjust," sometimes by eating the eggs of aggressive females to keep them from reproducing. Pruitt and Goodnight wrote in the journal *Nature*, "Many respected researchers have argued that group selection cannot lead to group adaptation except in clonal groups and that group selection theory is inefficient and bankrupt." The researchers proved that the group theory, like their spider colony, was doing just fine.

STRANGE MEDICAL CONDITION

SUBJECT: Natalie Adler of Melbourne, Australia

CONDITION: Unknown condition affecting the eyes

STORY: One morning in 2004, Adler, then 17, woke up and couldn't open her eyes. The condition lasted for three entire days...and then she could open her eyes again. Three days after that, she couldn't open her eyes, and three days after that, she could. Mind-bogglingly, that's been happening ever since. (For the three days her eyes are shut, Adler is legally blind, able to see only through a tiny slit in her left eye.) She went to doctors in both Australia and the United States, but none could determine what was causing her bizarre symptoms. At first, most of them told Adler the problem was in her head, but they don't anymore, because over the years, additional symptoms have appeared. They include headaches, nausea, stiffness in the neck and arms, and, most alarmingly, paralysis of Adler's stomach muscles, which has resulted in her having to be fed through tubes permanently inserted into her stomach. Adler has been seen by more than 40 specialists—and they still don't know what's going on. The latest guess: Adler is suffering from a previously unknown genetic disorder. Tests are ongoing.

Cause for ConCERN

Q: What is the world's largest machine?

A: The Large Hadron Collider, located on the Franco-Swiss border, is a 17-mile-long tunnel buried more than 500 feet below the surface where particles are charged and then accelerated to nearly light speed.

Q: What is the world's most dangerous machine?

A: The Large Hadron Collider. According to Stephen Hawking, Neil deGrasse Tyson, and several other concerned scientists, the experiments being carried out by CERN (the European Organization for Nuclear Research) could have catastrophic effects. Most worrisome is the "God particle," or Higgs boson, which was discovered at the Collider in 2012. In layman's terms, this is the subatomic particle that "gives mass to matter." Tyson warns that experiments with this high-energy, unstable particle could, if they go wrong, cause the Earth to "explode." (He actually used that word.) And if an experiment at the Large Hadron Collider goes *really* wrong, Hawking warns that it could "cause space and time to collapse."

FIVE FREAKY FACTS ABOUT...
THE NOBEL PRIZE

- The first Nobel Prize ceremony was in 1901, five years after the death of Alfred Nobel, who had mandated the prizes in his will. Nobel was a Swedish chemist who once blew up his own factory while working with nitroglycerin.

- Wangari Maathai of Kenya was the first African woman to win the Nobel Peace Prize, in 2004.

- John Enders cultivated polio in a test tube, and in 1954, he—not Jonas Salk, who developed the vaccine—got a Nobel Prize for his work.

- António Egas Moniz won the 1949 Nobel Prize in Medicine for developing the lobotomy.

- Gaston Ramon, a French veterinarian and biologist, was nominated for the Nobel Prize 155 times—but never won.

MORE MOVIE MAD SCIENTISTS

RE-ANIMATOR

In the mood for a really over-the-top splatterfest? *Re-Animator* has got the goods, a nasty—and nastily funny—flick in the *Frankenstein* mode (based on a story by creepmaster H. P. Lovecraft). In a nutshell: Testy medical student Herbert West (Jeffrey Combs) finds a formula to reanimate dead tissue, so he *does*. Hilarious and gory hijinks ensue. Not everyone's cup of tea, to be sure.

DR. STRANGELOVE, OR: HOW I LEARNED TO STOP WORRYING AND LOVE THE BOMB

Even if the entire film weren't already a brilliant black comedy about the end of the world by way of nuclear holocaust (and it *is*), this would still be worth seeing for the great Peter Sellers's portrayal of Dr. Strangelove, an expatriate Nazi scientist (very loosely modeled on Wernher von Braun). Strangelove is intensely weird, from the top of his toupéed head to the fingertips of his out-of-control (and self-homicidal) right hand. If actual nuclear scientists were anything like him, we'd all be glowing piles of ash.

EXPENDABLE ORGANS

GALLBLADDER

The gallbladder is a pear-shaped organ that sits just under the liver.

WHAT IT'S FOR: To store bile from the liver that's released as needed to digest fatty foods.

WHY WE CAN DO WITHOUT IT: Your gallbladder can get blocked by stones, creating a backup and infection that can be life-threatening.

SIDE EFFECTS OF REMOVAL? The bile constantly goes directly from the liver to the small intestine without the gallbladder acting as a gatekeeper. Rarely, patients experience frequent or constant diarrhea from that.

ANCIENT DATING TECHNIQUE

No, we're not talking about pickup lines. Since the 1700s, archaeologists have used what's called relative dating techniques to learn the approximate date of ancient artifacts. In most cases, they could only date an artifact relative to other finds. Biostratigraphy is the oldest technique, dating to the late 1700s. Simply put: new rock layers have been continuously created out of sediment on the Earth's surface for more than four billion years. Each layer contains fossils of the plants and animals that lived when that layer was created. For example: *Tyrannosaurus rex* fossils are found all over western North America—but only in rock layers corresponding to when T. rex existed (between 68 and 65 million years ago). So, if a new T. rex fossil is found, it—and fossils around it—can be relatively dated to that era. Biostratigraphy can be used to date artifacts from thousands to billions of years old, and was one of the key discoveries that led to Charles Darwin's theory of evolution. Today, it is used less often as more precise methods have been developed.

PSI-CHOLOGY

Is psi real—can people be "psychic"? Most reputable scientists will choose their words carefully and say that while telekinesis, clairvoyance, and the like may be possible, the existence of psi will not be accepted until 1) a scientist can orchestrate a controlled psi demonstration in a lab, and 2) other scientists can later re-create those results.

One reputable scientist set out to do just that. In 2010 Daryl Bem, a social psychologist at Cornell University, conducted several "retroactive precognition" tests on 1,000 college students. In one test, he showed them a series of two curtains on a screen, behind only one of which was an image. When the testers weren't told anything about the images, they chose the curtain with the image 50 percent of the time...an expected result. But when informed that there was an erotic picture behind one of the curtains, the college students chose the correct curtain 53 percent of the time, leading Bem to conclude that there must be something else at work to account for such a noticeable discrepancy. His resulting paper—"Feeling the Future: Experimental Evidence for Anomalous Retroactive Influences on Cognition and Affect"—passed the peer review process and was published in the esteemed *Journal of Personality and Social Psychology.*

So that means that the debate is over and psi is real, right? Not even close. Since the paper was published, no other scientist has been able to re-create Bem's results, and his testing methods have been called into question—as has the entire peer-review process. Even the *Journal of Personality and Social Psychology* ran a retraction. So for now, psi remains psi-fi.

 4 LESSER-KNOWN SCIENCES

1. **Mycology**—the study of fungi

2. **Ornithology**—the study of birds

3. **Limnology**—the study of inland waters

4. **Geobotany**—the study of plant distribution on Earth

A CURIOUS CURE

In 2005 Joyce Urch had a heart attack. The 74-year-old was rushed to Walgrave Hospital near her home in Coventry, England, where she underwent surgery and spent the next three days unconscious and near death. She finally woke up—and started shouting "I can see! I can see!" Urch had been blind for 25 years. "Then she leaned forward," said her husband, Eric, "and she just looked at me and said, 'Haven't you got old?' And I said, 'Wait 'til you have a look in the mirror.'" Just as doctors were unable to explain why she lost her sight so many years before (they thought it might be a genetic condition), they could give no explanation for its return. But the Urches didn't care. "When Joyce first went blind," Eric said, "everything seemed to fall away from us. This has given us both our lives back."

SNEAKY CORPORATIONS

The documentary film *An Inconvenient Truth* received a lot of attention and attracted huge audiences when it was released in May 2006. Narrated by Al Gore, the film argues that global warming caused by industrial pollution is altering Earth's climate and melting the polar ice caps, and will eventually flood major cities and leave the planet uninhabitable.

But shortly after the movie came out, "public service" commercials began appearing on TV, calling global warming a myth and claiming that carbon dioxide—a by-product of industrial pollution and automobile emissions (and the "villain" of the movie)—is actually not a pollutant at all, because "plants breathe it." They went on to say that industrial waste is not only harmless, it's essential to life.

So who made the "public service" ads? A think tank called the Competitive Enterprise Institute, whose members are almost exclusively oil and automobile companies, including Exxon, Arco, Ford, Texaco, and General Motors.

DUMB PREDICTIONS

"Everything that can be invented has been invented."

—CHARLES H. DUELL, COMMISSIONER, U.S. OFFICE OF PATENTS, 1899

INDECENT DUCKS

In 2005 biologists at Yale University were awarded a research grant to study the reproductive anatomy of the duck. Specifically, the researchers studied the unique corkscrew-like shape of the male duck's genitalia. After the study was showcased in Coburn's Wastebook and lambasted by cable news pundits, lead researcher Patricia Brennan defended her work. "This is basic science," she said. "The headlines reflect outrage that the study was about duck genitals, as if there is something inherently wrong or perverse with this line of research. Imagine if medical research drew the line at the belt! Genitalia, dear readers, are where the rubber meets the road, evolutionarily."

The cost of the study? $384,989.

MORE USED-LESS INVENTIONS

BIRD TRAP AND CAT FEEDER

PATENT NUMBER: 4,150,505

INVENTED IN: 1979

DESCRIPTION: For the crippled cat or the sparrow hater, this invention promises to "continuously supply neighborhood cats with plenty to eat." The trap lures birds with what appears to be an appealing perch and house, but once the feathered creature climbs through the entrance, it's caught in a pivoting plastic tube. The tube then lowers, dumping the bird into a wire mesh cage. Specifically designed for sparrows, the mesh is just big enough for the bird's head to poke through, which draws the cat's attention. A feeding frenzy presumably follows.

ANTI-EATING FACE MASK

PATENT NUMBER: 4,344,424

INVENTED IN: 1982

DESCRIPTION: The anti-eating device fits to the shape of a person's head with a series of flexible straps, rods, and hoops, while a gratelike mask covers the user's mouth from chin to nose, completely preventing the intake of food—except in liquid form. It's also fitted with a small padlock for insurance. Instead of locking the whole family out of the fridge, dieters can lock up their own mouths.

DANGEROUS APHRODISIACS

- Spanish fly, one of the most famous aphrodisiacs, is also one of the most dangerous. It has nothing to do with Spain or flies. It's really the dried, crushed remains of an insect known as the "blister beetle." Although it can constrict blood vessels, and thus may appear to be a sexual stimulant, it's actually a deadly poison. It can do irreparable damage to the kidneys.

- For thousands of years, people (especially in the Far East) have believed that by eating part of a powerful animal, a man can absorb its sexual vitality. This has led to the ingestion of such weird stuff as dried and powdered bear gallbladders, camel humps, and rhinoceros horns. (In fact, animal horns have been considered sexual stimulants for so long that the term "horny" became slang for "a need for sex.") It has also had a drastic effect on some endangered species. Poachers have slaughtered so many rhinos for their horns, which fetch up to $300,000 each, that only about 30,000 remain in the wild. And in North America, poachers have killed thousands of black bears to get their golf ball–sized gallbladders.

CAN'T SAY HE'S HEARTLESS

In August 2006, 55-year-old Louis Selo of London died while on vacation in Ireland. His body was examined at Beaumont Hospital in Dublin, where it was determined that Selo had died of a massive heart attack. The corpse was sent home to England, where another autopsy was performed (a second autopsy is customarily done when English citizens die out of the country). That operation went a bit differently: When the English doctors opened up Selo's chest, they discovered an extra heart and two extra lungs in a plastic bag inside him. An inquiry revealed that they were from an organ donor at the Irish hospital. Those organs were returned to the family of the donor, and an investigation to find out how they ended up in Mr. Selo was begun immediately.

How well do polar bears conserve heat? So well that they're barely detectable on thermal cameras.

SCIENCE FACTS THAT SOUND LIKE SCIENCE FICTION

I THINK IT CAN. In 2010 a swimming accident left 19-year-old Ian Burkhart paralyzed from the elbows down; he couldn't even move his fingers. Four years later, doctors at Ohio State University's Wexner Medical Center implanted a chip into Burkhart's brain in the area that controls movement. Then a cable was plugged into his head that was hooked up to a computer, which was hooked up to an electronic stimulation sleeve around his forearm. Burkhart thought very hard, and then...his fingers moved. The device completely bypassed his broken spinal cord, and for the first time since the accident, he could open and close his hand.

RESISTANCE IS FUTILE. The demilitarized zone between North and South Korea requires constant surveillance, but South Korea has 655,000 troops to North Korea's 1.2 million. So South Korea's military is turning to the Samsung Techwin SGR-A1 for help. This stationary robot is equipped with a camera and a high-speed machine gun. The camera scans the area and sends images to a control room; if there's trouble, the robot can be ordered to sound an alarm or fire 45-mm rounds. Multiple SGR-A1 are reported to be in place.

THAT HEALTHY RADIOACTIVE GLOW

Marie and Pierre Curie discovered radium in 1898, but it took decades of research for the long-term effects of radiation exposure to be understood. But in the interim, the general public regarded the stuff with almost superstitious awe. After all, it glowed with a beautiful phosphorescence!

Within a few years of its discovery, radium was—with no evidence whatsoever—being marketed as a restorative for youth and vitality. For that healthy glow, people used radium-laced toothpaste and face powder. Patients soaked in irradiated water to relieve rheumatism; heating pads loaded with radioactive ore soothed arthritis.

In particular, radium was reputed to cure sexual impotence. It shed its magical light in places where the sun don't shine as suppositories—and in the form of slender rods of radium-impregnated wax, to be inserted into the urethra. (Radium-dusted undergarments provided a less invasive option.) But then consumers and people who worked with radium started dying of cancer.

Now you can be grateful for those due-diligence regulations on the pharmaceutical industry. It may take years for innovative treatments to reach drugstores—but that's the trade-off for making sure that your heating pad doesn't give you cancer.

RANDOM ORIGIN:

GPS

You probably use GPS (Global Positioning System) on a regular basis, but do you know where it came from? Not long after the Soviet Union launched *Sputnik* in 1957, a team of American scientists monitoring the satellite's radio transmissions noticed that the frequency of its signal increased as it approached and decreased as it traveled away from them—a classic example of the Doppler effect. They realized they could use this information to pinpoint *Sputnik*'s precise location in space; conversely, if they knew the satellite's location, they could use it to determine their own location on Earth. This principle served as the basis for the U.S. military's NAVSTAR GPS, which became operational in 1993. The Americans intended to restrict the system to military use, but when the Soviets shot down a Korean Airlines flight in 1983 after it wandered into Soviet airspace, President Ronald Reagan announced that the system would be made available for public use.

SITH HAPPENS

Do the names Mikhail Lukin and Vladan Vuletic sound like *Star Wars* characters? They're actually Earth-based physicists (Lukin from Harvard, and Vuletic from MIT) who announced in 2014 that they found a way to "bounce" photons off each other. That's right—photons are light particles that have no mass. But when the physicists placed two photons into a special type of vacuum chamber, where they interacted with laser beams and a cloud of rubidium atoms cooled to near absolute zero, the light particles suddenly took on very different properties. "What we have done," said Lukin, "is create a special type of medium in which photons interact with each other so strongly that they begin to act as though they have mass, and they bind together to form molecules." Or, in other words, "It's not an inapt analogy to compare this to light sabers."

FIVE FREAKY FACTS ABOUT...
ALBERT EINSTEIN

- When he was little, Einstein didn't talk much, and the family's maid referred to him as "the dopey one."

- For years, he thought of his work in physics as a hobby and regarded himself as a failure...because he really wanted to be a concert violinist.

- As a teenager, he failed his college entrance exam. He tested well in physics and math, but his scores in other subjects, such as language and botany, were too low.

- While Einstein (who was Jewish) was living in Berlin, the Nazis burned his treatises and seized his belongings, including his violin. He fled Germany in 1932, just before Adolf Hitler came to power.

- He bequeathed the rights to his name to Hebrew University of Jerusalem, which trademarked it and made $10 million from licensing fees annually.

WHERE THE OCEAN MEETS THE SKY

Atmospheric gravity waves are a rare force of nature. They re-create the motion of the ocean, but in the sky. Waves of air move up and down as they roll through the atmosphere, fueled by buoyancy (the force that makes air rise) and gravity (which makes it fall). Meteorologist Tim Coleman explains: "Gravity is what keeps them going. If you push water up and then it plops back down, it creates waves. It's the same with air." Gravity waves begin when a stable layer of air is displaced by a draft (from a storm, for instance), causing air to ripple across the sky. It's similar to the rings that emanate outward when a stone is thrown into still water. Clouds develop high on the crest of each gravity wave and dissipate near the trough. So, to a person on the ground, gravity waves look like rows of clouds with clear sky between them. Their actual movement across the sky can really only be seen in time-lapse video.

MICROCHIP MAN

In July 1958, Jack St. Clair Kilby had been at his job just a few months at Texas Instruments. The entire plant shut down for a companywide two-week vacation, but Kilby hadn't earned a vacation yet. Virtually alone in the lab, he worked feverishly to come up with something to justify his employment.

At the time, the current that ran through electronic devices was conducted by transistors, which required workers to solder wires to hundreds, sometimes thousands, of microscopic gizmos, which—as you can imagine—was labor-intensive, expensive, and prone to errors.

Kilby managed to etch the entire circuit into a single sliver of germanium crystal. These "integrated circuits" made room-sized computers obsolete. And they were cheap enough to create a proliferation of electronic devices, including radios, microwaves, cell phones, VCRs, and TVs. Not only that, but Kilby and a colleague invented the handheld calculator—the first mass-market usage for the microchip.

Kilby snagged a Nobel Prize in Physics in 2000 for his work. When CNN asked him if he had any regrets about what his work produced, he answered, "Just one...electronic greeting cards that deliver annoying messages."

SULFONAMIDE TESTS

In 1942 doctors at the Ravensbrück concentration camp in northeastern Germany were ordered to test the effectiveness of a new kind of drug—an antibacterial called a sulfonamide. The Nazi government's goal was to reduce troop losses due to infection after injury, especially from gunshot wounds. To make conditions as true to life as possible, doctors at Ravensbrück were supplied with prisoner test subjects, most of them Polish women. Doctors cut long incisions into the women's calves, dabbed the wounds with virulent strains of bacteria, tied off the blood vessels at either end of the wounds—in order to simulate gunshot wounds—and then stitched up the incisions. Some of the women were given sulfonamide, some weren't; some were given small doses, some large. Their wounds were observed over the following weeks. Before the experiments ended in September 1943, 74 women had been subjected to the experiments. Eleven died. The rest suffered injuries to their legs that affected them for the rest of their lives.

THE HUMAN BODY AND THE EARTH'S CRUST

The 20 most prevalent chemical elements in the human body, and in the Earth's crust, by percentage.

HUMAN BODY

1. Oxygen: 65%
2. Carbon: 18%
3. Hydrogen: 10%
4. Nitrogen: 3%
5. Calcium: 1.5%
6. Phosphorus: 1%
7. Sulfur: 0.25%
8. Potassium: 0.2%
9. Chlorine: 0.15%
10. Sodium: 0.15%
11. Magnesium: 0.05%
12. Iron: 0.006%
13. Fluorine: 0.0037%
14. Zinc: 0.0032%
15. Silicon: 0.002%
16. Zirconium: 0.0006%
17. Rubidium: 0.00046%
18. Strontium: 0.00046%
19. Bromine: 0.00029%
20. Lead: 0.00017%

EARTH'S CRUST

1. Oxygen: 46.71%
2. Silicon: 27.69%
3. Aluminum: 8.07%
4. Iron: 5.05%
5. Calcium: 3.65%
6. Sodium: 2.75%
7. Potassium: 2.58%
8. Magnesium: 2.08%
9. Titanium: 0.62%
10. Hydrogen: 0.14%
11. Phosphorus: 0.13%
12. Carbon: 0.094%
13. Manganese: 0.09%
14. Sulfur: 0.052%
15. Barium: 0.05%
16. Chlorine: 0.045%
17. Chromium: 0.035%
18. Fluorine: 0.029%
19. Zirconium: 0.025%
20. Nickel: 0.019%

Ramming Speed

"Imagine traveling three kilometers in one second," said Michael Smart, an Australian aviation pioneer who has spent more than a decade developing a plane that's hypersonic (faster than five times the speed of sound) and can reach Mach 15. How fast is that? 11,509 mph! By contrast, the *Concorde* "only" traveled at about Mach 2. Smart's *scramjet* can achieve these speeds thanks to air being "rammed" through a super-efficient hypersonic engine so fast it creates intense heat that burns the fuel.

Initial tests have been promising, but there's still a lot of work to do. One holdup is that Smart's plane (which looks like a 1960s drawing of a futuristic rocket ship) requires a traditional jet engine to reach the speed that will allow the hypersonic engine to kick in. Another snag is finding an inexpensive material that won't burn up at such high speeds. But Smart is optimistic. He maintains that hypersonic planes—which will allow intercontinental travel in mere minutes, or even take people and equipment into space—will be flying through the upper atmosphere "in our children's lifetime."

TYPES OF COLOR BLINDNESS

You learned about color vision on page 108. Here are the conditions that can affect the way you see color.

1. PROTANOPIA. This is the lack of long-wavelength (red) cones, the effect being that reds look more like beiges and appear darker than they are; colors like violet and purple are seen as shades of blue because the red in them can't be seen.

2. DEUTERANOPIA. "Second blindness," or "red-green blindness," is the lack of the medium-wavelength, green-detecting cones; green and red appear identical to people with deuteranopia.

3. TRITANOPIA. "Third blindness" is the lack of short-wavelength cones (the blues). It makes blues and greens difficult to distinguish, and yellows can appear as shades of red.

4. BLUE CONE MONOCHROMACY. Also called "one color," only one type of cone, the blues, functions properly. A cone monochromat can see next to no color but otherwise has good vision in normal daylight.

5. ROD MONOCHROMACY. This is the condition of having only rods—and no functioning cones at all. It's the only condition for which the term "color blindness" is completely accurate—red monochromats can't see any color at all.

MYTHUNDERSTANDINGS

FIRESTARTER

MYTH: A lit cigarette carelessly tossed into a puddle of gasoline is likely to start a fire.

THE TRUTH: A mixture of gasoline vapor and air is highly explosive, but liquid gasoline will be wicked up by the cigarette paper and extinguish the burning ember.

GETTING SHOT

MYTH: When shot, a person is propelled backward.

THE TRUTH: Because a bullet is much lighter than a person, it doesn't have enough force to propel a person backward.

IS YOUR HEART IN THE RIGHT PLACE?

MYTH: The heart is on the left side of your chest.

THE TRUTH: It's actually more in the center of your body. What makes people think the heart is to the left is that the heart's left ventricle, a chamber that pumps blood, is larger than the right ventricle. This gives the heart its left-leaning shape, so that the heart intrudes farther into the left side of the body than to the right. It also gives the sensation of the heartbeat coming from left of center.

SMITHSONIAN BY THE NUMBERS

You haven't experienced the wonders of science until you've visited the Smithsonian Institution. Enjoy a numerical snapshot of this American must-see.

0

Number of firearms or dangerous weapons you're allowed to bring.

1

Exception to the above policy, per the Smithsonian website: "Kirpans (ceremonial knives) are religious articles of faith often worn by Sikhs. These knives are permitted in the museums as long as the blades are 2.5 inches or less in length."

2

Percentage of the Smithsonian Institution's holdings on display at any given time.

3

Number of one-cent stamps affixed to the first piece of mail flown across the Atlantic, which is housed in the Smithsonian's National Postal Museum.

4.8

Millions of botanical specimens housed by the Smithsonian's National Museum of Natural History.

19

Number of museums that make up the Smithsonian. Among others, these include the National Museum of the American Indian, the National Museum of African American History and Culture, and the Arthur M. Sackler Gallery (Asian art).

37.2
Weight, in tons, of a section of Route 66 delivered to the Hall of Transportation in the National Museum of American History for an exhibit.

45.52
Number of carats in the Hope Diamond at the Smithsonian Institution's National Museum of Natural History. It glows in the dark after exposure to UV rays and is semiconductive, too!

100,000
Amount of money, in British pounds sterling, that James Smithson originally willed upon his death in 1829. This eventually became the financial start of the Smithsonian.

154,000,000
Approximate number of objects, works of art, and specimens in the Smithsonian Institution.

CELEBRITY DISEASE

Jon Hamm, who starred in *Mad Men,* has vitiligo, a disease that causes the patient's skin to gradually lighten. (Michael Jackson famously suffered from it.) The cause is unknown, and there is no real treatment. Hamm said it only affects his hands, that it comes and goes, and that it was brought on by the stress of doing the award-winning show.

Green City:
CURITIBA

POPULATION: 3.5 million

HOW GREEN IS IT? Curitiba is the capital of the Parana state in Brazil, and despite facing severe poverty and overcrowding, it consistently wins recognition as one of the most beautiful, livable, and green cities in the world. In 1968 the city had less than 10 square feet of greenery per person, but careful urban planning—minimizing urban sprawl, planting trees, and protecting local forests—has turned that into 500 square feet for each inhabitant. Curitiba now boasts 16 parks, 14 forests, and more than 1,000 green public spaces. Curitiba is also internationally famous for its Bus Rapid Transport (BRT) system. Reliable and cheap, the BRT vehicles run as often as every 90 seconds in dedicated bus lanes. Eighty percent of the residents use the buses—that's more than two million riders a day. Also famous for its garbage disposal system, the city provides an alternative for low-income families who don't have garbage pickup: They can bring in bags of trash or recycling, and exchange them for bus tickets, food, school supplies, or toys. The result: A clean city where the poor live better and more than 70 percent of the waste is recycled.

DUMB PREDICTIONS

"Professor Goddard does not know the relation between action and reaction and the need to have something better than a vacuum against which to react. He seems to lack the basic knowledge ladled out daily in high schools."

—NEW YORK TIMES EDITORIAL, 1921
(about Robert Goddard's revolutionary rocket work)

STRANGE MEDICAL CONDITION

SUBJECT: Alexandra Allen of Mapleton, Utah

CONDITION: An allergy to water

STORY: When she was 12 years old, Alexandra was on vacation and went for a swim in a pool. She woke up in the middle of the night, most of her body covered in incredibly itchy hives. Her doctor initially diagnosed her as having a chlorine allergy and told her to stay out of pools. But it got worse as she got older—almost all physical contact with water was causing her skin to break out in hives and welts. Although the human body is about two-thirds water, it's still possible to be allergic to water (at least on the skin). Allen is one of a handful of cases of water allergy, or aquagenic urticaria, on record. Allen has to avoid being submerged in it, and can only take brief, cold showers. The main treatment? A topical application of capsaicin, the active ingredient in chili peppers.

JURASSIC FARTS

In 2012, two British scientists, Dave Wilkinson and Graeme Ruxton, revealed a startling discovery about dinosaur times. They were trying to figure out why the climate was so warm and wet when dinosaurs roamed the Earth. Their conclusion: sauropod farts. The scientists announced their theory in the journal *Current Biology*: "Our calculations suggest that sauropod dinosaurs could potentially have played a significant role in influencing climate through their methane emissions."

Sauropods were the biggest land animals to ever live on Earth. They had huge bodies, long necks, and puny heads. They first appeared in the late Triassic period (about 230 million years ago), had a heyday in the Jurassic period (200 to 145 million years ago), and went extinct at the end of the Cretaceous period (about 65 million years ago). Among the many sauropods were *Apatosaurus*, *Brachiosaurus*, and *Argentinosaurus*, which weighed up to 100 tons. (African elephants, the largest land animals alive today, average about 4 to 6 tons.)

Sauropods were herbivores. Scientists think they probably ate ferns, gingkoes, conifers, and similar plants. To maintain their size, they had to eat fast—they gulped their food without chewing it—and they had to eat a lot.

An 11-ton elephant can eat about 1,000 pounds of plant matter a day. Scientists say a 77-ton dinosaur would have had to eat at least four times that much, or 4,000 pounds of plants every day.

Scientists believe a sauropod's digestive system was a lot like a cow's. Because they don't chew, cows have four special stomachs to help digest their food. The long digestive process produces methane gas, and some of that gas comes out as burps and farts. Wilkinson and Ruxton did the math: given their size, giant sauropods could have produced more than 500 million tons of methane gas per year. "Our calculations suggest that these dinosaurs could have produced more methane than all modern sources—both natural and man-made—put together," said Wilkinson.

In the early 1800s, English scientist John Dalton proposed that each element—oxygen, gold, uranium, etc.—has a unique kind of atom. This began the modern era of atomic theory.

The Flowers of the Black Sea

In 2016 archaeologists were looking for the effects of climate change on ancient peoples, but instead they made the discovery of a lifetime: 44 well-preserved ships, some more than 1,000 years old, resting on the bottom of the Black Sea.

In most of the world's oceans, after a wooden ship sinks, sea creatures and barnacles feed on its hull until it's more reef than ship. But the Black Sea is different. It used to be a lake. At the end of the last ice age in 10,000 B.C., the Mediterranean rose and flooded it with saltwater, which settled on the seafloor in a layer so thick that there's very little sunlight and even less oxygen. Those conditions have created an eerie graveyard of well-preserved ships.

The crown jewel of the find is a nearly intact vessel from the Ottoman Empire that sank around 300 years ago. Archaeologists nicknamed it the *Flower of the Black Sea* because of its ornate carvings—even the ropes are still intact. Tethered robots were sent down to take high-resolution photos of the exteriors, but scientists won't explore the interiors until they can do so without damaging the wrecks. When they do, they could find all sorts of "flowers," or treasures. "You might find books, parchment, written documents," said archaeologist Brendan P. Foley. "Who knows how much of this stuff was being transported? But now we have the possibility of finding out. It's amazing."

TW iNS DAYS

Every summer for more than 40 years, identical and fraternal twins have been gathering in Twinsburg, Ohio, to celebrate twinhood at a festival called Twins Days. They can participate in the festival's parade, talent show, charity walk, and a variety of contests—from the best movie-themed costumes to the most similar- and dissimilar-looking twins.

The festival, which lasts three days, began in 1976 with 37 sets of twins; today, more than 2,000 pairs participate. It's been profiled on television programs like *Nova* and *That's Incredible* and is mentioned in the *Guinness Book of World Records* as the largest gathering of twins.

Here's where science comes in. Having so many sets of people with identical DNA in one place gives researchers an opportunity to study genetic and environmental influences on human beings. For example, they study the effects of medicine, cosmetics, and everyday products on twins. Even the FBI sends researchers to test facial recognition tools and biometric technology to try to distinguish between twins who have identical faces and DNA.

ACCIDENTAL DISCOVERY: PHOTOGRAPHY

The *camera obscura,* described by Leonardo da Vinci in the early 1500s, was widely used in the early 1800s—but not for photographs; the technology for photos didn't exist. People used the camera for tracing images instead, placing transparent paper over its glass plate.

In the 1830s, French artist Louis Daguerre began experimenting with recording a camera's images on light-sensitive photographic plates. By 1838 he'd found a way to capture an image using silver-coated sheets of copper.

However, the image was barely visible. He tried dozens of substances to see if they'd darken it, but nothing worked. Frustrated, Daguerre put the plate away in a cabinet filled with chemicals and moved on to other projects. A few days later, to his astonishment, the plate had darkened; the image was perfectly visible. One of the chemicals in the cabinet was almost certainly responsible...but which one?

He devised a method to find out. Each day he removed one chemical from the cabinet and put a fresh photographic plate in. If the plate darkened overnight, the chemical would be disqualified. If it didn't, he'd know he'd found the chemical. It seemed like a good idea, but even after *all* the chemicals had been removed, the plate continued to darken. Upon examining the cabinet closely, he noticed a few drops of mercury that had spilled from a broken thermometer onto one of the shelves.

Later experiments with mercury vapor proved that this substance was responsible. The daguerreotype's popularity paved the way for the development of photography.

That David's No Goliath

Michelangelo's *David* is one of the world's best-known statues, and represents the artist's ideal human form—even though critics have long wondered why the "ideal human form" would have such disproportionately small "private parts." *David* is 17 feet high and placed on a pedestal so admirers have to look up at him. But according to Pietro Bernabei, writing in the Italian journal *Il Giornale dell'Arte*, viewing David's face head-on, his blank expression changes to one of fear and worry. This makes sense—the statue depicts David just before his fight with the giant Goliath. And in a bit of dark humor, it explains the figure's "shrinkage": Male genitals typically recede when the body is under stress.

Superglue is so strong that a single square inch can lift a ton of weight.

FLORA FACTS

- Does Barbra Streisand smell better than former first lady Barbara Bush? Horticultural experts say yes, judging by the scents of the roses named after them.

- The substance in poison ivy that makes you itch is an oil called *urushiol.*

- It's possible to grow bananas in Iceland, in soil heated by underground hot springs.

- The trunk of the African baobab tree can reach a circumference of 110 feet.

- An olive tree can live for 2,000 years.

- What's in a name? The Venus flytrap feeds primarily on ants—not flies.

- As many as 179 species of tree can be found in a 2.5-acre area of rain forest.

- Where does vanilla come from? From a bean—the fruit of an orchid vine that is native to Mexico.

- There's a rose named for Whoopi Goldberg.

Generating a Regeneration Theory

"The invention of models to explain what nature is doing is the most creative thing scientists do," explains Michael Levin of Tufts University. So he decided to teach a computer how to invent its own model. Along with fellow computer scientist Daniel Lobo, Levin fed into a computer all of the data they could find on flatworms. They were trying to solve a mystery that so far has eluded biologists: how flatworms can regenerate into more flatworms when they're sliced into little pieces.

The computer spent three days crunching the numbers and processing the data until—all on its own—it came up with the flatworm's "core genetic network" that finally explained how the invertebrate could regenerate. "None of us could have come up with this model," Levin told *Popular Mechanics* in 2015. "We have failed to do so after over a century of effort." His ultimate goal is to teach a computer how to find the solution to an even bigger mystery that has baffled scientists for centuries: how to cure cancer.

7 NATURAL WONDERS OF THE WORLD

THE GRAND CANYON was created by millions of years of wind and water erosion from the Colorado River. The rocks of the canyon walls range from 250 million years old at the top to more than 2 billion years old at the bottom.

PARICUTÍN VOLCANO erupted out of a Mexican cornfield on February 20, 1943. Located about 200 miles west of Mexico City, Paricutín grew to 10,400 feet in just nine years, making it the fastest-growing volcano in recorded history. Its lava destroyed two villages and hundreds of homes, but caused no fatalities.

THE HARBOR OF RIO DE JANEIRO in Brazil was first seen by Portuguese explorers on January 1, 1502. The Portuguese thought they had reached the mouth of an immense river and named their find River of January— Rio de Janeiro. The spectacular harbor's landmarks include Sugarloaf Mountain and Corcovado Peak.

THE NORTHERN LIGHTS, also called the auora borealis, occur when solar particles from the Sun collide with gases in Earth's atmosphere. The energy created by the collision is emitted as photons (light particles)—and we see lights dancing across the sky.

VICTORIA FALLS, the world's largest waterfall, lies between Zambia and Zimbabwe in Africa, where the Zambezi River plummets 420 feet over a cliff. Although Scottish missionary David Livingstone named it after the Queen of England, native Africans call it Mosi-oa-Tunya ("The Smoke that Thunders"), because the falling water makes thunderous clouds of spray.

THE HIMALAYAS, the highest mountain range in the world, formed about 60 million years ago. India (at that time a separate continent) rapidly moved northward and collided with Asia, and the crash produced these amazing mountains. The famous Mt. Everest stands above the other peaks at 29,035 feet, making it the tallest on the planet.

THE GREAT BARRIER REEF off the coast of Queensland, Australia, is the world's largest coral reef, with an estimated 1,500 species of fish and 350 types of coral. It is over 1,400 miles long and can be seen from space!

TECHNOLOGICAL DIFFICULTIES

Man created technology in his own
image...which is why it's so weird.

WORST APP EVER

There are thousands of apps for Apple's iPhone, but none drew more complaints than the "Baby Shaker": a video game in which the player shakes the iPhone until a virtual baby stops crying (then two red Xs appear over its eyes). The app was only available for download for two days in 2009 before Apple removed it. The company explained that it should have been rejected before it was added, but someone must have "missed it." Alex Talbot, the app's designer, admitted, "Yes, the Baby Shaker was a bad idea."

HOW TO GET ON THE NEWS

Say, what's that suspicious-looking device? It's the "Suspicious Looking Device"! This real product you can purchase is a darkly humorous response to the increased fears of terrorism in recent years. What is the SLD? It's a red metal box with dotted lights, a small screen, a buzzer, and whirring motor. What does it do? Nothing. It's just supposed to *appear* suspicious. So if you want to see your name in the headlines, just place the SLD in front of your local police station.

Can You Dig It?

On page 43, we told you about the fascinating applications of nano-gold, particles of gold so small that they measure in nanometers. (A million nanometers could line up single file across the head of a pin!) But where does nano-gold come from?

Currently, scientists must make their own nano-gold by dissolving larger pieces of gold and growing nanocrystals. That may soon change. A research team has found nano-gold in western Australian clay. The area's salty, acidic water dissolves gold deposits in the clay and redeposits them in masses of gold nanoparticles. But finding extractable deposits isn't easy. "Gold nanoparticles are transparent and effectively invisible," explains lead scientist Dr. Rob Hough. Why bother? Invisible gold—just like the kind you can see—is worth $1,500 an ounce and is projected to rise to $15,000 per ounce by 2020. All you have to do is find it.

MILK AND MICROBES

Louis Pasteur wasn't much of a student in his youth, but he developed an interest in the process of fermentation when he became a dean at Lille University in France in 1854. In 1865 he decided that there must be some sugar-fed organism in wine and beer that was busy reproducing and giving off gas, a microorganism so tiny that it was invisible to the naked eye. This came to be known as "germ theory," and it led to Pasteur being dubbed "the father of microbiology."

Soon everyone realized that bacteria, although you can't see them, are everywhere and in everything. And they come in battalions, not as single little sneaky guys. You've got a bunch of them right now inside you: without them you can't digest anything because they live, love, and work in your gut. They're also what makes yeast work. Bacteriology was a shiny new science, and in its wake came a number of life-enhancing developments.

Heat, Pasteur knew, kills bacteria; by experimentation, he found that heating milk or another food to 161.6°F for 15 seconds and then cooling it quickly killed the bacteria. That way, disease-causing bacteria can't be passed from the cow to the human who drinks the milk.

By the time Pasteur died in 1895, his name was everywhere. Almost all the milk sold in the Unites States and most of Europe was being routinely pasteurized. Result? A massive drop in the incidence of typhoid, diphtheria, dysentery, and tuberculosis. Today, almost all milk is pasteurized.

Pasteur was also the man behind immunization. A large number of diseases are caused by an invasion of harmful bacteria. If the body's own army of antibodies can't get rid of them, you're in trouble. "Why not send in reinforcements?" reasoned Pasteur. "Strengthen resistance by making the antibodies multiply. Theoretically, that would prevent diseases from developing." And that is exactly what immunization does.

Mixed-Up Geology

"Wait, volcanoes are real? I thought they were made up."
—a high school freshman

UNSUNG SPACE TRAVELERS

FRUIT FLIES

The first travelers in space were not humans, dogs, or apes, but plain, everyday fruit flies. In 1947, at the beginning of the space race, the United States launched a V-2 rocket carrying seeds and fruit flies in an effort to study the effects of radiation beyond Earth's atmosphere. The rocket went up 68 miles to the edge of outer space. Then the capsule detached, the parachute engaged, and the capsule fell back to Earth. The bugs and seeds survived the journey.

IBERIAN RIBBED NEWTS

In 1985 Soviet scientists operated on 10 newts, amputating one of their front limbs and an eye lens before launching them into space. But why? The Soviets wanted to know if the missing limbs would regenerate in zero gravity in the same way they regenerated on Earth. In fact, the newts healed significantly faster when in space than similarly amputated control groups back on Earth.

ROUNDWORMS

When the space shuttle *Columbia* disintegrated over Texas in 2003, its seven crew members died. So did the silkworms, garden orb spiders, carpenter bees, Japanese killifish, and harvester ants that had been on the shuttle with them as part of various experiments. The only known survivors were roundworms called nematodes that were found intact in the debris.

More "Science" Museums

PAPER HOUSE MUSEUM

Location: Rockport, Massachusetts

Background: It took 20 years and 100,000 newspapers for the Stenman family to build this tiny house. It's only one room, but everything is made entirely of newspapers. The walls are made of 215 layers of newspaper. Fireplace, chairs, tables, desk—you guessed it—all out of newspaper.

MUSEUM OF BEVERAGE CONTAINERS

Location: Goodlettsville, Tennessee

Background: In 1973 Tom Bates started to pick up empty cans on the walk home from school. Eventually he opened a museum, and it was even listed in the *Guinness Book of World Records* for its collection of 36,000 cans and bottles. Among the treasures: soft drinks with names like Zing, Zippy, and Zitz, a can of soda for pets, and camouflage cans produced for the U.S. Army during World War II. Unfortunately, it has since closed, and we're crushed.

Ancient Art

Petroglyphs are a remarkable reminder of past civilizations. Indigenous people made these rock carvings by scratching or chipping at dark rock to unveil its lighter surface underneath. Some fascinating examples exist in northeastern Arizona, where Anasazi Indians made petroglyphs that date to at least A.D. 1000. Despite being exposed to hundreds of years of harsh sunlight, desert winds, and monsoon rains, many engravings remain intact. The art includes squiggles, stick figures, scenes, and complex geometric patterns. Their meanings generally confound archaeologists, but many believe the Anasazi—who had no other written records—used the pictures to communicate important information about their village and its residents. Scientists speculate that the carvings may be maps, prayers, or even warnings that say, "Hey, foreigners, this is *our* village."

Some experts think that some of the petroglyphs function as solar calendars. The Anasazi relied on the angle of the sun's light to tell them of impending season changes. When sunlight intersects with certain spirals and other carvings on the rocks, it acts like a sundial, letting them know when the days would become longer or shorter and when they could start planting crops.

• • • • • • • • • • • • • • • • • • • •

Although the human race is about 200,000 years old, one out of every ten humans ever born is alive right now.

Skinner's Box

B. F. Skinner was a behavioral psychologist like John B. Watson (see page 192). His early experiments with rats and pigeons showed that they could learn quickly with rewards, otherwise known as "positive reinforcement." His success encouraged him to try the process on a human being. And who better than his own daughter, Deborah?

Skinner had performed his animal experiments in a controlled chamber that he called the "Skinner Box," so when Deborah was born in 1944, he created a similar item: a glass-enclosed combination of crib, playpen, and diaper-changing station into which warm air circulated so that Deborah could sleep and play comfortably in her diapers, without the confinement of clothing or blankets. She seemed to enjoy the arrangement, and her mother liked doing less laundry. But then came the urban legend.

It all began in 1945 when Skinner wrote an article about his "air crib" for *Ladies' Home Journal.* Unfortunately, the photo that ran with the article depicted the baby in a smaller, portable box that was different from her usual box and was captioned "Baby in a Box." Eventually, the legend grew to horrific proportions, to the point that a 2004 book titled *Opening Skinner's Box* claimed that Skinner kept his baby in a Skinner Box while running experiments that left Deborah so deranged that she later killed herself.

The urban legend made baby Deborah famous—and the book made grown-up Deborah furious. In fact, she'd grown up to be a well-adjusted adult. She countered the stories about her father's experiments with a newspaper article entitled "I Was Not a Lab Rat," in which she angrily refuted the book's claims.

> "Science may have found a cure for most evils; but it has found no remedy for the worst of them all—the apathy of human beings."
>
> **—Helen Keller**

DIRTY TRIX

The Trix Rabbit is trying to brainwash your children. And so is Cap'n Crunch, Toucan Sam, and dozens of other cereal box cartoon characters. That's what a team of researchers at Cornell University concluded in 2014. They discovered that on cereals marketed to kids, the characters' eyes are shifted down 9.6 degrees...directly at kid level in the cereal aisle. That way, the eyes "follow kids around" (much like a museum painting). Test subjects were shown one of two similar boxes of Trix—one with the silly rabbit looking down, the other altered so he's looking straight ahead. The subjects who were met by the rabbit's gaze "increased feelings of connection to the brand by 28 percent." The study concluded with two recommendations:

- If you are a cereal company looking to market healthy cereals to kids, use spokes-characters that make eye contact with children.

- If you are a parent who does not want your kids to go "cuckoo for Cocoa Puffs," avoid taking them down the cereal aisle.

When a Man Loves a Bird

In honor of Tesla's obsession with the number 3,
page 333 tells a little-known story about him...

Nikola Tesla wasn't much of a ladies' man, and rumor has it he died a virgin. But that doesn't mean that the Serbian-born scientist—who invented the X-ray machine and a lot of other cool things—wasn't a romantic. He did find love in his life...with a pigeon. And according to Tesla, it was reciprocal. He adored all of the birds that he fed each day in New York City's Bryant Park, but there was one white female that was truly special. "I loved that pigeon as a man loves a woman," he said.

By that point—the 1920s—Tesla had become a bitter old recluse with little money to his name...or even much of a name. He was living alone in a hotel room, isolating

himself from a world that, despite his unequaled genius and countless contributions, had shunned him. Feud after feud and watching others take credit for his work was too much to bear, so he found solace in feeding the birds.

But then one day, tragedy struck: Tesla found his favorite pigeon with a broken wing and leg. He rushed her back to his room and spent weeks nursing her back to health. He even built a special harness for her to sleep in without hurting her wing. After the bird healed, she returned to the scientist's hotel room often and even flew to him when he called her.

Sadly, the pigeon wasn't long for this world. One evening she landed next to Tesla, who sensed that the bird was near death. The old scientist cared for her well into the night. And then something amazing happened: "Two powerful beams of light," he later wrote, emanated from the pigeon's eyes. "A powerful, dazzling, blinding light, a light more intense than I had ever produced by the most powerful lamps in my laboratory." Then she died.

Tesla could never explain what the light was, but he knew at that moment that he was done with science, so he hung up his lab coat for good. "As long as I had her," he lamented, "there was a purpose to my life."

HOW P2P WORKS

- Peer-to-peer file sharing, or simply P2P, refers to a specific type of Internet file-sharing network. It allows people with the appropriate programs to make their computers part of an entire file-sharing network, including its bandwidth (meaning how much data can be transferred) and storage capabilities. That means that as more people enter into a P2P network, they automatically increase the network's ability to handle more information.

- This stands against the more conventional client/server systems in which a finite number of servers—computers that "serve" many users in varying ways—can become slower as more and more people use them. (If you have an e-mail account with Yahoo! or Google, for example, you're the client, and you use their servers to send and receive e-mails.)

- The concept of a P2P system was actually part of how the early Internet worked: In 1969 ARPANET, the "Grandfather of the Internet," connected computers at UCLA, Stanford, UC Santa Barbara, and the University of Utah in what was basically a P2P system students could use to access the different schools' computers and files. Each computer was, therefore, both server and client.

STRANGE LAWSUIT

THE PLAINTIFF: Carl Sagan, world-famous astronomer

THE DEFENDANT: Apple Computer, Inc.

THE LAWSUIT: Late in 1993, computer designers at Apple codenamed a new computer model "Carl Sagan." Traditionally, this is an honor—"You pick a name of someone you respect," explained one employee. "And the code is only used while the computer is being developed. It never makes it out of the company." Nonetheless, Sagan's lawyers complained that the code was "an illegal usurpation of his name for commercial purposes" and demanded that it be changed. So Apple designers changed it to BHA. When Sagan heard that it stood for "Butt-Head Astronomer," he sued, contending that "Butt-Head" is "defamatory on its face."

THE VERDICT: Case dismissed.

Q: What is the strongest creature known to exist?

A: The gonorrhea bacterium (*Neisseria gonorhoeae*). It can pull 100,000 times its own weight.

ANATOMY OF A HICCUP

- A hiccup occurs when a stimulus causes an involuntary contraction of the diaphragm, the muscle separating the lungs from the abdomen. The contraction makes the sufferer take a quick breath, causing the glottis (located in the voice box) to close, which makes the "hic" sound.

- Most common causes: too much alcohol, spicy food, cold water, carbonated drinks, indigestion, or asthma. They can also be caused by liver or kidney problems, abdominal surgery, or a brain tumor.

- Fetuses hiccup in the womb.

- Folk cures: eat peanut butter, eat wasabi, drink vinegar, eat lingonberry jam, drink a glass of water while urinating.

- Unlike other body reflexes (coughs, sneezes, vomiting), hiccups serve no useful purpose.

- The word "hiccup" may come from the French *hocquet*, which was used to describe the sound of a hiccup. The earliest known version in English is *hicket*, dating from the 1500s.

- Hiccup lore: In ancient Greece, a bad case of the hiccups meant an enemy was talking about you. To get rid of them one had to guess the enemy's name. The Scots thought holding your left thumb (or your chin) with your right hand while listening to someone singing a hymn would stop the hiccups.

- Some forms of *encephalitis* (swelling of the brain) can cause hiccuping. During the encephalitis pandemics of the 1920s, several cities reported cases of mass hiccuping.

- Technical term for hiccups: a diaphragmatic spasm, or *singultus*.

"Scientific" Theory: HOW TO BUILD A SCORPION

For thousands of years, people believed that living things could grow out of nonliving things. Aristotle, for example, believed that oysters grew out of slime, and eels out of mud. Known as spontaneous generation, this was treated as fact for ages—and even persisted until fairly modern times. In the 17th century, for example, Flemish chemist Jan Baptista van Helmont, one of the most respected scientists of any era (he was the first to show that air was composed of different substances, and even coined the word "gas"), believed that you could create animals by following simple recipes. Van Helmont's notes, for example, contain a recipe for making mice: Put some wheat on a dirty cloth inside an open container, let it sit for 21 days and—voilà!—mice will be created. Another: Put some basil between two bricks in sunlight. Then...scorpions. It wasn't until the mid-1800s that scientific progress finally saw spontaneous generation spontaneously combusted for good.

More Pop (Culture) Science

YOU CAN HEAR ME NOW

Scientists are hard at work making a universal translator a reality. Microsoft's Skype Translator provides real-time translation between languages as people speak, and went live in 2016. The app was demonstrated at a conference in 2014: An English-speaking man in California chatted with a German-speaking colleague via Skype, with verbal and textlike translations appearing on-screen. The app still has some bugs to work out (it is Microsoft, after all), but expect universal translators to be a common sight before the end of the decade.

BLAST 'EM!

Dr. Evil (from the Austin Powers movies) had one simple request: "sharks with frickin' laser beams attached to their heads." Now his dream is a reality (minus the sharks). In 2014 the U.S. Navy demonstrated its LaWS (Laser Weapons System). The prototype—installed on the USS *Ponce*—has a targeting, tracking, and firing system that is operated via a "video game–like controller." According to an ominous military press release, the weapons officer "manages the laser's power to accomplish a range of effects against a threat, from disabling to complete destruction." (Add maniacal laughter here.)

DUMB PREDICTIONS

"You want to have consistent and uniform muscle development across all of your muscles? It can't be done. It's just a fact of life. You just have to accept inconsistent muscle development as an unalterable condition of weight training."

—COMMENT TO ARTHUR JONES
(INVENTOR OF NAUTILUS EXERCISE EQUIPMENT)

AMAZING AMBER

Amber has been valued as a precious gem since ancient times, but today jewelers have to fight off paleontologists to get their hands on these precious gems. Because of how it is formed, amber droplets can hold a wealth of information for paleontologists. Amber starts out as tree resin; when exposed to the air, tree resin usually dries out and crumbles. If the resin is protected from oxygen (like if it its buried under clay and silt), it will, over millions of years, fossilize into amber.

Before it hardens, the sticky resin sometimes traps bits from its environment. Feathers, animal hairs, plants, and insects become fossilized along with the resin. Called inclusions, these remnants of ancient life, so fragile that they might not leave a fossil trace, are perfectly preserved inside bulbs of amber.

Fast-forward to the 1800s, when clay was mined in Sayreville, New Jersey. Large pits were dug to harvest the clay, and amber was found. But serious interest in Sayreville's amber didn't take off until the early 1990s, when a fossil hunter dug into an abandoned pit and discovered amber that held an insect.

While New Jersey doesn't have the largest deposits of amber in the United States, it has the only substantial North American deposits that date back to the Cretaceous period (65–135 million years ago). At that time, dinosaurs were still around, and flowering plants and modern insects began to emerge. Sayreville's amber and its inclusions contain perfectly preserved specimens of life from the Cretaceous. Paleontologists have found New Jersey amber with inclusions that date back as far as 95 million years, and discovered more than 100 species of insects and plants trapped inside the petrified resin. Sayreville's amber has housed the world's oldest:

- Ant—the first proof that ants lived during the Cretaceous.

- Mosquito with a mouth tough enough to bite a dinosaur!

- Mushroom, *Archaeomarasmius leggetti.* It's 90 to 94 million years old.

- Bee, *Trigona prisca.* It flew for the last time about 65 to 80 million years ago.

You Do *WHAT* in Your Car?

These unique inventions will allow you to multitask while you drive.

INVENTION: Integrated Passenger Seat and Toilet apparatus (1988; patent no. 4,785,483)
INVENTOR: Paul H. Wise; Tucson, AZ
DETAILS: This inventor figured that when you've got to go, you've got to go, so he designed a way to conceal a toilet under a passenger seat with a swiveling chair. It includes a built-in privacy curtain, a foot pedal flush system, a holding tank beneath the car, and an electric water pump.

INVENTION: Flushable Vehicle Spittoon
(1991; patent no. 4,989,275)
INVENTOR: Dan L. Fain; Chancellor, AL
DETAILS: Some drivers love to chew tobacco or munch sunflower seeds as they head down the road—but there's always the problem of where to spit. The Flushable Vehicle Spittoon attaches by Velcro to your door or dashboard, to be ready wherever and whenever you need it. Gravity drains the waste through a funnel into a tube that empties into the great outdoors. A separate line attaches to the windshield wiper fluid container, providing an extra flush should gravity fail to do the job. Unfortunate pedestrians walking nearby will just have to step lively to avoid being sprayed by any discharge.

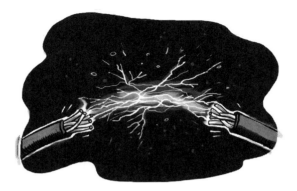

Edison the Executioner

By the early 20th century, electricity was spreading to homes and businesses across America. The preferred method was AC, or alternating current, electricity being promoted by George Westinghouse. One guy not too happy about that: Thomas Edison, who came up with DC, or direct current, electricity. Edison staged publicity stunts to prove that his method was superior.

Edison went about promoting DC power, and showing that it was safe and effective, in a bizarre way—he'd use AC power to publicly electrocute animals. Edison killed cats, horses, an orangutan, and once helped the state of New York execute a convicted ax murderer. Did Edison prove once and for all the greatness of DC power by using it to kill a slew of powerful beasts? Not exactly—AC remained the industry standard.

BALLOON BOMBS AWAAAY!

On May 5, 1945, a group of picnickers in Oregon fell victim to one of the oddest weapons used in World War II. The party found a 32-foot balloon in the woods. When they tried to move it, it exploded, killing six of them—the only fatalities of World War II to occur on American soil. In 1944, feeling intense pressure from American air raids, Japan came up with a seemingly brilliant way of striking back. Their planes couldn't fly all the way to the States, so they started sending balloons made of rubberized silk, each carrying an explosive device. The balloons were supposed to ride high-altitude winds across the Pacific and come down to wreak havoc on the American heartland. The U.S. Air Force estimates that Japan launched 9,000 weaponized balloons between November 1944 and April 1945. About 1,000 of those actually made it to the United States, but they inflicted only minor damage. Designed to be a weapon of mass terror, Japan's balloon bomb campaign was rather ineffective. Had the Japanese added "germ bombs" to their balloons, casualties might have been immense. It's likely that they balked at such a step for fear the United States would retaliate with their own germ weapons.

Rubin's Cubed

Vera Rubin, the astronomer whose work provided the first real evidence of dark matter (see page 89), was one of the few bright female stars in a male-dominated field. As she found this anomaly quite strange, she spent much of her life advocating for female scientists until her death in 2016. Along the way, she came up with "Three Basic Assumptions." Here they are:

1. There is no problem in science that can be solved by a man that cannot be solved by a woman.
2. Worldwide, half of all brains are in women.
3. We all need permission to do science, but, for reasons that are deeply ingrained in history, this permission is more often given to men than to women.

Proving that progress does happen, in 2001, Princeton University (which Rubin was unable to attend because of her gender) appointed Shirley M. Tilghman as the first woman president in the school's 255-year history.

LEECH THERAPY

Leeches are worms that live in freshwater and latch onto victims to suck up their blood. When these creatures bite, they secrete an anticoagulant called *hirudin* that prevents the victim's blood from clotting. That makes it easier for the leeches to feed. After being bitten by a leech, a person can bleed for hours. For thousands of years, doctors used leeches for bloodletting. They thought bloodletting could cure everything from headaches to hemorrhoids. By the mid-1800s, it became clear that bloodletting didn't work, and leeches went down in history as a medical mistake.

RETURN OF THE LEECHES: In the 1980s, leeches made a comeback. Plastic surgeon Joseph Upton had reattached the severed ear of a five-year-old boy—an amazing medical feat at the time—but the tissues in the ear were dying. Upton rounded up some leeches, attached them to the boy's ear... and it healed. Why? When a body part is reattached, blood may pool and clot, blocking circulation. Leeches suck up the extra blood, which prevents swelling. Their saliva releases chemicals that numb pain, fight infection, and calm inflammation. The fresh blood that flows to the damaged tissue helps it to heal and to produce new growth.

STRANGE SCIENCE

PURPLE TO THE PEOPLE

In 1856 an 18-year-old chemist named Sir William Henry Perkin was trying to create a new medicine for malaria. When he mixed together several chemicals, he found not a cure...but an unusually beautiful purple dye. Up until that point, purple dye was acquired through costly, drawn-out processes—it was extracted from animals (see page 135). Fabrics made from these dyes were available only to the rich. Perkin's family saw the commercial potential for his accidental discovery. They named the dye *mauveine*, or *mauve*. Within a year they opened their first dye factory, bringing inexpensive and easy-to-make purple fabric to regular people like us.

URINE GOOD HANDS

Dr. Savely Yurkovsky of Chappaqua, New York, says he knows a cure for the deadly disease SARS. Victims must collect some of their own infected saliva, mucus, and urine, mix it with a little water, and drink it. The potion, Yurkovsky says in his book, *Biological, Chemical and Nuclear Warfare: Protecting Yourself And Your Loved Ones,* will trigger the immune system to go after the disease. Other revelations in Yurkovsky's book: "Poison can be your best friend," "Carcinogens can protect you against cancer," and "Toxic chemicals can extend lifespan and enhance immunity."

In 2014 Neil deGrasse Tyson interviewed God* for his podcast.

*or, at least, the person whose twitter handle is @TheTweetofGod

---

Genius School

There's a school in New Jersey that has no classes, no tests, no degree programs, and...it's free. The catch? You have to be a genius to get in. The school was started in 1930 by noted educator Dr. Abraham Flexner. He wanted to give genius types a place to lose themselves in the world of ideas, a place where they would not have to worry about things like cooking or cleaning. The school is called the Institute for Advanced Study, and it's located at 1 Einstein Drive in Princeton. At first, it was just a school for mathematics. Later other departments were added: Historical Studies, Natural Sciences, and Social Sciences.

The geniuses invited to stretch their brains at IAS have included Albert Einstein, John von Neumann (father of game theory), J. Robert Oppenheimer (father of the atomic bomb), Kurt Gödel (called "the most important logician of our times"), and Hetty Goldman (archaeologist and the first female genius at IAS).

RANDOM ORIGIN:
Organ Transplants

On June 17, 1950, an Illinois surgeon named Dr. Richard Lawler removed a kidney from a donor who'd been declared brain dead moments earlier and transplanted it into a 49-year-old woman named Ruth Tucker. Kidney dialysis had only recently been invented and was not yet widely available; for most people, failing kidneys were still a death sentence. A transplant wasn't very promising either— doctors still hadn't figured out how to stop the human body from rejecting transplanted organs. Lawler went ahead with the surgery anyway. The transplanted kidney did fail several months after the surgery, but not before taking strain off of Tucker's remaining kidney, which began functioning normally again. Tucker lived another five years before dying of a *heart* complication; Dr. Lawler never performed another transplant. "I just wanted to get it started," he explained years later. (The first successful organ transplant, between identical twins for whom rejection was not an issue, followed in 1954.)

MANIMALS!

PIGGY BLOODY PIGGY

Jeffrey Platt, director of the Mayo Clinic Transplantation Biology Program in Minnesota, performed human stem cell injections into fetal pigs, and now has a group of pigs that have pig blood cells *and* human blood cells running through their veins. But it gets weirder: Some of the blood cells are both. Their DNA contains both human and pig genes. Platt hopes the work might lead to pigs being raised for their human blood and organs, but there are several hurdles, including the fact that some pig viruses can be passed on to humans.

MORE HOLLYWOOD PHYSICS

FICTION: Falls from great heights are easy to walk away from.

SCIENCE: Stunt players seem to do it all the time, but falls from as "high" as three feet (like falling out of bed) can cause serious injuries. Just remember that the farther a body falls, the harder it lands.

FICTION: A spaceship needs to bank when turning to compensate for the effects of centrifugal force.

SCIENCE: In Earth's atmosphere, aircraft have to bank to create a pressure difference on the two wings in order to produce the turning force. Despite what you've seen small spacecraft do in the *Star Wars* movies, the only forces necessary for them to change direction in the vacuum of space are the rockets that propel them.

FICTION: When a space station explodes, it makes a deafening noise.

SCIENCE: There's no air in outer space to transmit sound waves, so those big explosions you hear in *Star Wars* are pure fiction.

FICTION: Laser beams are visible.

SCIENCE: Though the end point is visible, the beam itself is only visible when reflected by the fine particles in mist or smoke.

THE SCIENCE OF SINGING SANDS

The first part of the story is on page 255.

Dunes that "sing" need a special recipe of sand, moisture, wind, and movement. In particular, the dunes must be created out of grains of sand that have been blown over long distances, making them unusually smooth and round. All the grains must be similar in size, and the dunes free of foreign particles. Humidity and moisture also affect the sound—too much moisture and the sand goes silent because the grains can't move. But the dune must have some rainfall so that its inner grains stay a little damp.

When a dune creates sound, its outer layer of sand (which must be a few feet thick) is dry from the Sun, but its inner core can be wet. Wind then pushes sand grains to the top of the dune and they accumulate until the angle of the slope reaches a tipping point of about 35 degrees. That causes an avalanche of sand grains to fall, creating friction and producing the loud bass tone similar to one produced by a stringed instrument.

BRAINPUT

As modern drivers, we all know the feeling of wanting to pay attention to the road but getting distracted by equally important activities like watching cat videos on our cell phones or eating a cheeseburger. The solution? Brain probes and robots, of course. Scientists at MIT, Indiana University, and Tufts University are working on something called "Brainput" to help us multitask more safely and effectively, because let's face it: We're never going to learn to actually pay attention to what we're doing.

The system works like this: You attach two probes to your forehead, and, using what is purportedly science but sounds an awful lot like diabolical voodoo, the probes sense when your attention is drifting and notify a robot to pitch in and help you out (e.g., your car drives itself for a minute so you can focus on licking a dollop of errant mustard off of the steering wheel). If the concept of robots reading your mind causes you to repeatedly scream and then run way, don't worry just yet: The technology has only been tested in a very basic form, in which the robot helps the user navigate a maze. Brainput's application in real-life situations, like the driving-while-texting-and-noshing mentioned above, is still but a distant dream/nightmare.

DR. YESTERYEAR

- Doctors in ancient India closed wounds with the pincers of giant ants.

- The world's first recorded tonsillectomy was performed in the year 1000 B.C.

- Sixteenth-century French doctors prescribed chocolate as a treatment for venereal disease.

- Leprosy is the oldest documented infection—first described in Egypt in 1350 B.C.

- Among the "treasures" found in King Tut's tomb: several vials of pimple cream.

- Acne treatment, circa A.D. 350: "wipe pimples with a cloth while watching a falling star."

- In medieval Japan, dentists extracted teeth with their hands.

- The Hunza people of Kashmir (India and Pakistan) have a cancer rate of zero. Some scientists link it to the apricot seeds they eat.

- In the Middle Ages, Europeans "cured" muscle pains by drinking powdered gold.

- Doctors in the 1700s prescribed ladybugs as a cure for measles; they were to be ground up and eaten.

- Between 1873 and 1880, some U.S. doctors gave patients transfusions of milk instead of blood.

- During World War I, raw garlic juice was applied to wounds to prevent infection.

- People in ancient China would swing their arms to cure a headache.

BOMBING MARS

One method to quickly make Mars more Earthlike (and thus more hospitable to humans) was suggested by aerospace engineer Robert Zubrin in his 1996 book *The Case for Mars*. The plan: astronauts would attach a nuclear thermal rocket engine to a 10-billion-ton asteroid (kind of like in the movie *Armageddon*). Controlled remotely from Earth, the asteroid would hit Mars with the force of 70 hydrogen bombs. The impact would raise the Martian temperature 3°F, which would melt a trillion tons of ice. This would add CO_2 to the atmosphere, triggering the greenhouse effect and melting the caps even more. One asteroid-bomb per year over 50 years could make up to 25 percent of the Martian surface habitable (temperature-wise, anyway). And scientists could then send their algae rockets to the planet's new seas.

Sadly, all those nukes would soak Mars in toxic radiation (so we'd better be sure there's no life there), and humans would have to wear air tanks for hundreds of years anyway. But no matter how we get there, Zubrin writes, the time to begin the journey is now. "We need a central overriding purpose to drive our space program forward. At this point in history, that focus can only be the human exploration and settlement of Mars."

Romancing the Stone

Many gem scholars attribute the tradition of birthstones to the jeweled "breastplate of Aaron" described in the Bible. The breastplate was a ceremonial religious garment worn by Aaron, the brother of Moses; it was set with 12 gemstones representing the 12 tribes of Israel and perhaps, say folklorists, the 12 months of the year.

Around that same time, the Assyrians began assigning gemstones to each region of the zodiac according to a color system that they believed controlled its power. Each stone had its own distinct magical, protective, and curative qualities that corresponded with the attributes of the astrological sign. Over time the stones came to be associated more with calendar months than astrological signs.

The custom spread to other cultures—including Arabic, Jewish, Hindu, Polish, and Russian, each of which modified the list of birthstones. Over the centuries, other changes and substitutions were made: sometimes

accidentally by scribes, sometimes by royalty who didn't like their birthstones, and sometimes according to fashion and availability.

In 1912 the American National Association of Jewelers came up with the Traditional Birthstone List, a standardized list that combined contemporary trends with all the birthstone lists from the 15th to the 20th centuries. A few years later, it was revised and renamed the Modern Birthstone List. The association hoped the modern list would eliminate confusion among jewelers.

Did it work? Not entirely. The old lists didn't go away, so there are still variations in jeweler's lists. And those aren't the only lists, either. There's a Mystical Birthstone list that's based on ancient Tibetan culture, an Ayurvedic list originating from the 1,000-year-old system of Indian medicine, a zodiac list, and a planetary list, to name just a few.

It takes an estimated 40,000 years for a photon to travel from the core of the Sun to its surface...and then only about 8 minutes to travel to Earth.

THE RAT-HEAD EXPERIMENT

In 1924 Carney Landis, a graduate psychology student at the University of Minnesota, designed an experiment to determine whether there is a basic underlying human facial expression for any given emotion. In the school's lab, Landis drew black lines on the faces of several volunteers (fellow grad students) to more easily track the movements of their facial muscles. He then photographed their faces as he exposed them to stimuli meant to evoke specific emotional responses, including exposing them to the smell of ammonia, having them stick their hands into a bucket of live frogs, and having them watch pornographic films. Then came the final experiment: Landis gave each of the students (one at a time, with no one else present) a live rat and a large, sharp knife—and instructed each student to decapitate the rat. Two-thirds agreed to do it, and actually cut off the rats' heads. The other third refused, so Landis decapitated the rats for them, while taking photographs of their (disgusted) faces. Conclusion: Landis discovered no universal facial expressions, but did find that most test subjects will do whatever they're told to do. (Our conclusion: Landis liked to kill rats.)

DUMB
PREDICTIONS

"Louis Pasteur's theory of germs is ridiculous fiction."

—PIERRE PACHET, PROFESSOR OF PHYSIOLOGY AT TOULOUSE, 1872

HOW TO MAKE ICE

How do they maintain the ice in rinks, especially in warm-weather places like Florida? We caught up with Ken Friedenberger, Director of Facility Operations for the St. Petersburg Times Forum, home of the Tampa Bay Lightning hockey team, for the rundown:

- Two layers of sand and gravel mixture form the foundation of the ice. The two layers and the precise mixture, Friedenberger said, prevent it from freezing into permafrost (perpetually frozen soil), which would "eventually crack the piping and turn it into a big mess, which would look like spaghetti."

- "The piping" he refers to is perhaps the most important part of the rink. Five to ten miles of it run under and through a massive concrete slab that sits on the base. A liquid similar to antifreeze is cooled by massive air conditioning units to below freezing and pumped through the piping, making the temperature of the concrete slab below freezing, too.

- Water is hosed onto the concrete and allowed to freeze in a very thin layer. When it's frozen, more water is added and allowed to freeze, another layer is added... and the process is repeated until there are 24 layers of ice, each one from $\frac{3}{4}$ of an inch to a full inch thick.

- When all of this is finished, the ice surface temperature hovers between 22°F and 26°F. And because of the constantly cooled concrete below, the temperature inside the stadium stays in the 60s or 70s even when the air temperature outside is in the 90s.

- The lines, circles, and spots are painted on before each game, and four to five new layers of ice are frozen over them to protect them.

- A Zamboni machine smooths out the ice before a game—and it's time for the opening faceoff.

8 ELEMENTS DISCOVERED BY THE
ANCIENT GREEKS AND ROMANS

1. Antimony
2. Copper
3. Gold
4. Lead

5. Mercury
6. Silver
7. Sulfur
8. Tin

ALBERT EINSTEIN SAYS...

"I never think of the future. It comes soon enough."

"Common sense is the set of prejudices acquired by age eighteen."

"Nationalism is an infantile disease. It is the measles of mankind."

"To punish me for my contempt for authority, Fate made me an authority myself."

"Why is it that nobody understands me, and everybody likes me?"

"With fame I become more and more stupid, which of course is a very common phenomenon."

"A life directed chiefly toward fulfillment of personal desires sooner or later *always* leads to bitter disappointment."

"My political ideal is that of democracy. Let every man be respected as an individual, and no man idolized."

"Science without religion is lame, religion without science is blind."

"I am a deeply religious nonbeliever...This is a somewhat new kind of religion."

"Try not to become a man of success, but rather, a man of value."

THE TIPLER TIME MACHINE

Frank Tipler has been fascinated with time travel since he was five years old. In 1974 he designed a time machine that he hoped would work in real life.

In his design, a person would take off in a spaceship and arrive at a cylinder rotating in space. Tipler believed that if the cylinder had enough mass and was rotating fast enough, it would work like an artificial black hole and have the power to warp time. After orbiting the cylinder the spaceship would go backward in time and it would be the past when the ship returned to Earth. The design had a few problems: to generate a black hole, the cylinder would have to be *infinitely* long. Once within the vortex created by the fake black hole, your ship would not be able to generate enough velocity to escape. So...you'd be stuck there. But not for long: Your ship would be crushed, and you'd be dead.

Lounge Lizards

The Gila monster, native to Mexico and the U.S. Southwest, is North America's most venomous lizard. Thankfully, it's so slow that it poses little threat to humans. And now it might even help us. In 2012 a Swedish scientist named Karolina Skibicka created a synthetic version of a substance found in the Gila monster's saliva called *exendin-4* that "affects the reward and motivation regions of the brain." In clinical tests, rats that were treated with the drug had reduced cravings for chocolate. "This is both an unknown and quite unexpected effect," explains Skibicka, adding that "our decision to eat is linked to the same mechanisms in the brain which control addictive behaviors." Her hope is that one day the lizard spit drug will help reduce cravings in not just people with a sweet tooth, but alcoholics as well.

WHO'S THE *RAREST* OF THEM ALL?

A *circumhorizontal arc* (aka, "fire rainbow") is the world's least common natural atmospheric condition, but it's actually neither a rainbow nor a fire. Although the phenomenon appears as rainbow-colored clouds sporting wisps that look like flames, it is produced by ice crystals, not warmth, and conditions have to be perfect for one to form. Start with cirrus clouds more than 20,000 feet high. Then add the Sun also high in the sky, at least 58 degrees above the horizon. The clouds must contain hexagonal ice crystals just the right thickness and aligned horizontally with a flat face pointed at the ground. Similar to a prism, light enters through the vertical side face and exits through the flat bottom, producing an arc of colors that lights up the cloud. Some of these anomalies cover hundreds of square miles and last for more than an hour. However, due to the specific conditions they require, the arcs are impossible to view in latitudes below 55 degrees south or above 55 degrees north. Sorry, Canada!

MAD DOGS AND A DEADLY DISEASE

Louis Pasteur worked on a number of vaccines, including one for treating rabies. Human beings get it by being bitten or even licked by infected animals, mostly dogs, who drool and look mad—not angry, but crazy. The bad news is that no one has ever been known to recover from rabies. However, thanks to Pasteur, you can prevent it from taking hold.

Pasteur was so sure that his rabies vaccine would work that he was ready to deliberately inoculate himself with rabies in order to demonstrate his discovery. Before that could happen, a nine-year-old boy named Joseph Meister arrived in Pasteur's laboratory. Joseph had been bitten two days earlier by a rabid dog. So, fortunately for Pasteur, he had a guinea pig other than himself. The treatment involved a ten-day course of injections; Joseph survived—and so did Pasteur's reputation.

FIVE FREAKY FACTS ABOUT...
FRO-YO

- Frozen yogurt isn't just regular yogurt that's been frozen. If you put yogurt into a soft-serve machine, you'll end up with a milky slush. (And if you put *that* into the freezer...you'll get a weird brick of white stuff.)

- The ingredient that chemically gives fro-yo an ice-cream-like consistency is sugar—lots of sugar, as much as ice cream has. The science: sugar molecules block ice crystals from forming.

- Frozen yogurt contains a bunch of processed dairy products and dairy by-products, such as pasteurized nonfat milk, pasteurized buttermilk, whey, dry milk, and milk protein isolate—plus carrageenan, an extract from seaweed.

- Most commercial frozen yogurts include live yogurt cultures.

- Self-serve frozen yogurt parlors that charge by weight have become one of the most heavily franchised businesses. They've got low overhead, and the yogurt is sold at a markup estimated at about 500 percent.

MUSICAL AILMENTS

FIDDLER'S NECK

The name might sound silly, but according to a study of regular violin and viola players by Dr. Thilo Gambichler of Oldchurch Hospital in London, the friction of the instrument's base against the left side of the neck (for right-handed players) can cause lesions, severe inflammation, and cysts. What's worse, said the study, published in the British medical journal *BMC Dermatology*, it causes lichenification—the development of a patch of thick, leathery skin on the neck, giving it a "bark-like" appearance.

GUITAR NIPPLE

A similar report issued in the United States cited three female classical guitarists who suffered from traumatic mastitis—swelling of the breast and nipple area—due to prolonged friction from the instrument's body. The condition can strike male guitarists, too.

BAGPIPER'S FUNGUS

Recent medical reports have detailed the dangers of playing Scotland's national instrument. Bagpipes are traditionally made of sheepskin coated with a molasses-like substance called treacle. That, the report said, is a perfect breeding ground for various fungi, such as *aspergillus* and *cryptococcus*. Bagpipers can inadvertently inhale fungal spores, which doctors say can lead to deadly lung (and even brain) diseases.

Tesla vs. Edison

The early 20th century's most talented inventors—Nikola Tesla and Thomas Edison—were rivals. When Edison died in 1931, the only dissenting voice came from Tesla, who had worked with Edison. "He had no hobby, cared for no sort of amusement of any kind and lived in utter disregard of the most elementary rules of hygiene," Telsa wrote in the *New York Times*. It gets worse:

> His method was inefficient in the extreme, for an immense ground had to be covered to get anything at all unless blind chance intervened and, at first, I was almost a sorry witness of his doings, knowing that just a little theory and calculation would have saved him 90 percent of the labor. But he had a veritable contempt for book learning and mathematical knowledge.

In fact, this rivalry might be the reason neither man was ever awarded a Nobel Prize. The pair came close in 1915 when they were the favorites to win the Nobel Prize in Physics for their work on X-rays. Instead, it went to British scientists William Henry Bragg and his son W. L. Bragg "for their use of X-rays to determine the structure of crystals." Tesla and Edison would have likely been awarded the prize...if not for their very bitter public feud.

HEART HISTORY

- In a cave in Pindal, Spain, there's a surprisingly well-drawn wall painting of an elephant made some 50,000 years ago. In the chest area is a red mark that some people argue is the creature's ear, a handprint, or a mistake on the artwork. But others say it's the animal's heart.

- Ancient Egyptians knew about hearts (which they removed and stored in jars alongside their wrapped, entombed dead).

- Early Chinese medical practitioners knew it as part of the circulatory system.

- The centuries-old sacred Hindu text, the *Atharvaveda*, included chants for heart health.

- The Greeks were also aware of a beating heart in the body, but thought it was the home of the soul. The brain, many of them believed, was what kept us alive.

- By about the 16th century, humankind had a pretty good idea what the heart looked like and what exactly it did, thanks to increasingly sophisticated scientific methods and public acceptance of dissection for the sake of information.

- The Romans were the first to propose that the heart was a person's emotional center, so it didn't take much to make the connection between the heart and love. The physical reactions we have when we fall in love—that heady rush, the flip-flop in the midsection, the increase in pulse— probably led to that belief.

- The ancients thought that there was a vein that ran from the fourth finger of the left hand directly to the heart (there isn't), which is why we wear wedding rings on the left "ring finger."

The Language Of Cat Whiskers

Whiskers can be used to communicate. Here are some tips for deciphering what a cat's whiskers are saying:

- A calm, resting, or friendly cat holds his whiskers out to the sides.

- An alert, curious, or excited cat's whiskers point upward.

- Backward-pointing whiskers often indicate that a cat feels defensive or is angry. So, you... human...the one with the kitty shampoo and bath supplies in hand: Back off!

STRANGE MEDICAL CONDITION

SUBJECT: Graham Harrison of Exeter, England

CONDITION: Cotard's syndrome

STORY: In 2004 Harrison, then 48, went to a doctor with an odd complaint: he was dead. He explained that he'd attempted to commit suicide several months earlier (he took an electrical appliance with him into a bath), and while he seemed to have survived, he was convinced he had, in fact, killed his brain. Harrison recognized that he could walk and talk, but he was still convinced his brain had ceased to function. His rationale: he had lost all sensation, he said, including the ability to smell, to taste, and to feel pleasure. He was so convinced he was dead that he saw no point in eating. (His family had to make sure he ate food and took his medications.) Finally in 2013, after nine years of suffering with the symptoms, doctors diagnosed Harrison's condition as Cotard's syndrome, also known as "walking corpse syndrome," an extremely rare psychological disorder, the cause of which is unknown. People with Cotard's sincerely believe they are, basically, zombies. After the diagnosis, Harrison was able to return to the land of the living: "I don't feel that brain-dead any more," he told *New Scientist* magazine. "Things just feel a bit bizarre sometimes."

In the course of being treated, Harrison underwent brain scans, and they were, according to his doctors, surprising: Harrison's brain showed a level of activity similar to someone in a vegetative state. "I've been analyzing scans for 15 years," said Dr. Steven Laureys. "I've never seen anyone who was on his feet, who was interacting with people, with such an abnormal scan result."

AN ANIMAL ODDITY

In the early 1800s, French naturalist Georges Cuvier was studying a female argonaut, a small kind of octopus, when he discovered a strange parasitic worm inside its abdomen. He named the worm hectocotylus, "one hundred cups," for the many suckerlike structures on it. It wasn't until the 20th century that biologists discovered that it was the male argonaut's penis, and it was supposed to be inside the female. Nearly all cephalopods— the class of marine animals that includes octopuses, squids, and cuttlefish—have this specialized tentacle that detaches once the job is complete. With the argonaut and similar species, it can actually break off before contact with the female...and swim over to her to do its business.

A Singer Who Butchered Science

Superstar Bette Midler recorded a very popular song that makes casual reference to a basic scientific phenomenon...and got the facts completely wrong. Midler's 1988 ballad "Wind Beneath My Wings" was featured in the movie *Beaches* and won a Grammy for Song of the Year. The sentiment of the song is simple: the singer thanks a friend for always supporting her, for helping her metaphorically fly—the "wind beneath her wings." The problem is that this is not how flight works. For the song's narrator to "fly higher than an eagle"—or at all— wind would have to be moving *above* the wings, not below. But in any case, calling someone the "wind beneath my wings" sounds like a completely different kind of wind.

WEIRD ENERGY:
ALGAE

One of the main obstacles in developing alternate energy sources is that alternatives are more expensive than oil. The fact that algae-based energy could be as cheap as petroleum might make it one of the more viable long-term options. Currently, crude oil is pumped out of seabeds, where it was created from heat and pressure, transforming algae and other microorganisms over the course of millions of years. Scientists at the U.S. Department of Energy's Pacific Northwest National Laboratory in Washington state have devised a way to re-create and speed up the process of turning algae into oil. If they scale it up to mass production, the lab estimates that it could sell this biofuel for about $2 a gallon.

Other benefits: the technology that converts algae into petroleum creates fertilizer that can be used to make more energy in the form of natural gas. The big negative: real estate. To produce enough energy to meet just 17 percent of the country's current needs, we'd need an area the size of South Carolina for algae production.

ACCIDENTAL DISCOVERY: PENICILLIN

Dr. Alexander Fleming had spent most of 1928 working in a cramped laboratory in a London hospital. While working on the influenza virus, he had filled his lab with culture dishes containing staphylococci bacteria. Exhausted from too many late nights, Fleming decided to take a break, giving strict instructions to his assistants on how to care for his specimens. On his return, however, Fleming was annoyed to find that someone had left a window open the previous night. The result? A foreign mold had flown in through the window and contaminated the culture dishes. A devastated Fleming went to dispose of the dishes when something caught his eye—moldy patches were growing all over the plates, but there were rings of clear space around them where there were no bacteria. Looking closer, Fleming saw that the bacteria closest to these clear rings were shriveling or dissolving. The astute doctor began experimenting with this mold that appeared to eat up bacteria. After years of research he was able to extract from it a drug—penicillin—that has saved millions of lives. And it was all because someone forgot to close the window.

The Cow Egg Man

In April 2008, a team of scientists at Newcastle University in England extracted an unfertilized egg cell from a cow, removed its nucleus—where most of a cell's DNA resides—and replaced it with the nucleus of a cell taken from another animal. They then gave the egg a tiny electric shock, which "activated" it, meaning that the inserted DNA began to do its work, and the cell started dividing. In other words, it was alive. The DNA they inserted into the cow egg was taken from a human skin cell. The Newcastle scientists had successfully cloned a human-animal hybrid, possibly for the first time in history. However, the cells stopped dividing after about three days. But the team hopes to repeat the experiment and get an egg to keep dividing for about six days—at which time it should begin creating embryonic stem cells, the "building block" cells found in embryos that go on to become more than 200 different types of cells in the body. The cells would consist of 99.9 percent human DNA and only 0.1 percent cow DNA. If successful, the procedure would allow scientists to skirt around laws forbidding or restricting the use of "normal" human embryos for stem-cell production.

"I WAS AT HOME, ASLEEP!"

Somnambulism, or sleepwalking, affects millions of people. Yet the rarest and most unusual type of this disorder involves not only sleepwalking but elaborate, murderous actions—"sleep killing." Take the 1987 case of Kenneth Parks of Toronto. He drove 14 miles to his in-laws' house, where he stabbed his mother-in-law to death and choked his father-in-law into unconsciousness. Afterward, he woke up and drove to a police station. He turned himself in, his hands and clothes covered with blood, still unaware of what he had done. A jury found Parks not guilty because he had no conscious control over his behavior.

A 7.0 earthquake is 900 times more powerful than a 5.0 earthquake.

ANCIENT SOAPMAKING

According to some accounts, around 1000 B.C., Romans performed many animal sacrifices to the gods on Mount Sapo. The fat from the animals mixed with the ashes of the sacrificial fires. Over time, this mixture of fat and alkali flowed down to the Tiber River and accumulated in the clay soils. Women washing clothing there found that the clay seemed to help get things cleaner. Whether or not this story is true, experts say Mount Sapo is the origin of the word *soap*.

However, a recipe for soap was discovered on Sumerian clay tablets dating back to 2500 B.C. During excavations of ancient Babylon, archaeologists uncovered clay cylinders containing a soaplike substance that were around 5,000 years old. The Phoenicians were making soap around 600 B.C., and Roman historian Pliny the Elder recorded a soap recipe of goat tallow and wood ashes in the first century. A soap factory complete with finished bars was found in the ruins of Pompeii.

In Spain and Italy, soapmaking did not become an established business until about the 7th century. France followed in the 13th century and England a century after that. Southern Europeans made soap using olive oil. Northern Europeans used the fat from animals, including fish oils.

In most places, soap was a luxury item because it was so difficult to manufacture. And it was often so heavily taxed that it was beyond the budgets of most people. Furthermore, bathing was out of fashion for many centuries, being considered sinful, even unhealthy. But when Louis Pasteur proved in the mid-1800s that cleanliness cuts down on disease, bathing and the use of soap for personal hygiene became a more accepted practice.

If you could tap the energy released by an average-sized hurricane, it would be enough to satisfy all U.S. energy needs for six months.

YOU AREN'T "YOU"

The average adult human body is made up of more than 30 trillion cells, all of them descendants of the one original *zygote*—the egg cell from your mother that was fertilized by sperm from your father. The average body also contains microbes: organisms that enter from the environment, such as the bacteria that line the intestines and aid in digestion. These beneficial microbes colonize our bodies during and shortly after birth, and they stay with us our entire lives. In fact, the average adult human body contains as many colonizing microbes as native human cells.

Oops!

City officials in Nottingham, England, spent more than £1 million (about $1.5 million) installing solar-powered parking meters on city streets after reading reports that the meters saved a fortune in maintenance costs in Mediterranean countries. The only problem: Mediterranean countries get a lot of sun...and England doesn't, not even in summer. More than 25 percent of the parking meters went out of commission, allowing hundreds of motorists to park for free.

Meteorologists' Jargon

bomb: unforeseen weather; also called a gremlin

cold low: a weather system that consists of a low-pressure area riding above a mass of cold air at the earth's surface, bounded by slowly swirling bands of cloud and precipitation in the upper atmosphere

Colorado hooker: a storm that originates over the eastern plains of Colorado and moves across the central Plains and Mississippi Valley; the storm taps moisture from the Gulf of Mexico and pulls cold air down from Canada as it travels its hooked, or curved, path

dead clouds: cumulus clouds that blot out the sun but are usually incapable of generating precipitation

helicity: a measure of the potential of a small area of the atmosphere to spin rapidly, forming a tornado; expressed as a number, it is used to describe the danger of tornadoes forming

hot box: a localized storm area squared off on a weather map for which a meteorologist is likely to issue severe storm warnings; helicity is often closely monitored within hot boxes

severe clear: not a cloud in the sky

sucker hole: a brief period of clear weather that lasts until just after the good weather forecast has been issued or until a pilot flying without instruments takes off; only a sucker amends the forecast to reflect the good weather, and only a sucker flies without instruments when such conditions occur

Texas gully washer: rain intense enough to flood gullies

TTTC: weather that is "too tough to call"

INSTANT DRUNKENNESS– REVERSING PILLS

Imagine being able to get rip-roaring drunk, raise hell for a few hours, and then pop a few pills and sober up quicker than you can say, "I'll be glad to walk on that line, officer!" Well, we're soon going to find out, because scientists appear to have unlocked the key to countering the intoxicating effects of alcohol: enzymes. A team of researchers led by Yunfeng Lu, a UCLA professor of chemical and biomolecular engineering, and Cheng Ji, a professor of biochemical and molecular biology at USC, has devised a way to package enzymes inside a nanoscale polymer shell. In non-egghead speak, they found a way to put chemicals inside your body that can change what your body does. Anyway, they tested these tiny capsules on drunk mice and found that the enzymes caused their blood alcohol levels to drop quickly and significantly. Professor Lu stated that down the road he can envision an alcohol prophylactic or an antidote that could be taken orally after someone becomes inebriated. Nothing bad could possibly come of that, right?

"PSYCHIC DRIVING" PROCEDURE

In the early 1950s, Scottish-born psychiatrist Dr. Ewen Cameron developed what he believed was a cure for schizophrenia, and in 1953 began testing it on patients at the Allan Memorial Institute in Montreal. He called it "psychic driving." The treatment: Patients were drugged into unconsciousness with powerful sedatives, had earphones placed on their heads, and were subjected to repeated messages, such as "People like you" or "You have confidence in yourself," over and over...and over... for days, weeks, and, in some cases, months. Over a decade, Cameron subjected hundreds of people to "psychic driving." Not a single person is believed to have been cured or even helped by the treatment—and many were quite likely made worse off. At the same time Cameron was performing these experiments, he was taking part in the CIA's notorious MKUltra "mind control" program, involving, among other things, dosing unwitting subjects with LSD. That's why Cameron's clinic was in Canada—it would have been illegal to do such things in the United States.

SNAPSHOT OF SCIENCE
MOON MAN
July 20, 1969

Most people can tell you who was the first man to set foot on the Moon. It was U.S. astronaut Neil Armstrong on July 20, 1969. As he lowered himself onto the lunar surface, Armstrong immortalized the moment with his famous words, "That's one small step for a man, one giant leap for mankind." When it comes to iconic photos of a man on the Moon, however, one picture stands out from the rest. It isn't of Armstrong. Rather, he was the photographer. He snapped the photo of fellow Moon man Buzz Aldrin, the second man to step on the surface.

Armstrong took the famous photo of Aldrin saluting the American flag firmly planted in the dusty surface, with a dunelike mountain rising up behind him into the ink-black cosmos. In the reflection of Aldrin's lowered helmet visor is the image of Armstrong and the lunar lander *Eagle*.

DUMB PREDICTIONS

"I think there is
a world market
for maybe five
computers."

—Thomas Watson,
chairman of IBM, 1943

PUTTING THE "BYE" IN ANTIBIOTICS

In 1918 the Spanish flu killed more people than died in the entirety of World War I. Between 50 million and 100 million people succumbed, roughly 5 percent of the world's entire population at the time. So it was nice when scientists discovered antibiotics shortly thereafter! Some of the most effective disease killers, antibiotics (such as penicillin and amoxicillin) have helped prevent infections like the flu from turning into devastating global pandemics. It's really too bad that they *just* stopped working.

Antibiotics work by killing bacteria, while usually leaving safe, healthy human cells mostly unharmed (as opposed to more extreme disease-and-healthy tissue-killing treatments like radiation or chemotherapy). They're basically poison for bacteria, but over the past 70 years, the bacteria have developed a tolerance.

Mutation guarantees that a small percentage of bacteria would eventually become antibiotic-resistant. But when antibiotics have been added to everything from hand soap to animal feed (which gets passed on to the humans who eat those animals), mutation and evolution accelerated. While many strains died, some of those that didn't became supergerms that are resistant to traditional drugs.

MOHS HARDNESS SCALE

In 1812 German mineralogist Friedrich Mohs developed a scale to measure the hardness of different materials. He based it on ten commonly used minerals. Here they are, from softest to hardest:

1. Talc: used in talcum powder
2. Gypsum: used in plaster of Paris
3. Calcite: found in limestone
4. Fluorite: used in the manufacture of steel and glass
5. Apatite: found in tooth enamel and most types of rock
6. Feldspar: found in granite
7. Quartz: quartz crystals
8. Topaz: gemstone
9. Corundum: found in sapphires and used in abrasives
10. Diamond

The scale works by determining the hardest material on the list that another material can scratch. For example, a human fingernail can scratch gypsum (#2) but can't scratch calcite (#3). Human fingernails therefore have a hardness of 2.5 on the Mohs scale. Window glass can scratch apatite (#5) but not feldspar (#6)—and has a hardness of 5.5. Because the relative hardnesses of the minerals are not mathematically proportional, the Mohs scale is very simplistic. Topaz is twice as hard as quartz, but diamond in nearly four times as hard as corundum. The scale is still used today as a relative hardness measurement.

Used-Less Inventions

"HIGH FIVE" SIMULATOR

PATENT NUMBER: 5,356,330

INVENTED IN: 1994

DESCRIPTION: Essentially a spring-loaded arm mounted on a wall, the "High Five" Simulator is always ready for a good slap. A fake hand attached to a forearm piece is connected to a lower arm section with an elbow joint for pivoting. When the hand is struck, the raised arm bends backward briefly before returning to the ready position. This invention is perfect for the lonely or excessive high-fiver.

TOO MANY MUMMIES

Pharaohs weren't the only ancient Egyptians who were mummified—nearly everyone in Egyptian society who could afford it had it done. By the end of the seventh century A.D., the country contained an estimated 500 million mummies. Egyptians from the 1100s onward thought of them more like a natural resource than as the bodies of distant relatives, and treated them as such.

For over 400 years, mummies were one of Egypt's largest export industries. By 1600 you could buy a pound of mummy powder in Scotland for about 8 shillings. As early as 1100, Arabs and Christians ground them up for use as medicine, which was often rubbed into wounds, mixed into food, or stirred into tea.

By the 1600s, medicinal mummy use began to decline, as many doctors started questioning the practice. "Not only does this wretched drug do no good to the sick," the French surgeon Ambrose Pare wrote, "...but it causes them great pain in their stomach, gives them evil smelling breath, and brings on serious vomiting which is more likely to stir up the blood and worsen hemorrhaging than to stop it." He recommended using mummies as fish bait.

By the 1800s, mummies were imported only as curiosities, where it was fashionable to unwrap them during dinner parties.

Mummies were also one of the first sources of recycled paper: During one 19th-century rag shortage (in the days when paper was made from *cloth* fibers, not wood fibers), one Canadian paper manufacturer imported Egyptian mummies as a literal source of raw materials. He unwrapped the cloth and made it into sturdy brown paper, which he sold to butchers and grocers for wrapping food. The scheme died out after only a few months, when employees in charge of unwrapping the paper began coming down with cholera.

4 MAJOR TYPES OF VERTEBRATE TISSUE

1. Epithelial tissue
2. Connective tissue
3. Muscle tissue
4. Nerve tissue

WEIRD SCIENCE NEWS

THE MYSTERIES OF LIFE

In 2013 Japanese scientists reported findings of a study in which they observed 108 pairs of sea slugs having sex. The researchers reported that after mating, a male sea slug wanders off—and its penis falls off...and another penis grows back in just 24 hours. (This can happen at least three times!) The findings also showed that the sea slugs being studied—*Goniobranchus reticulata*, gathered from the East China Sea—are hermaphrodites, meaning they have both female and male sex organs and can impregnate another sea slug and be impregnated simultaneously.

WHO CARES?

German psychologist Thomas Goetz of the University of Konstanz concluded a study in 2013 that showed there aren't four types of boredom, as previously believed—but five. (Goetz also conducted the study that concluded there were four types of boredom.) The five types: indifferent boredom; calibrating boredom; searching boredom; reactant boredom; and, the latest, apathetic boredom. "We did not expect this type of boredom at all," Goetz said of his latest boring discovery.

WEB WEAVER

The World Wide Web might not have happened if Tim Berners-Lee had kept better office files. In 1980 the man who would create the ubiquitous "www" that sits in that topmost bar on your computer screen was into a six-month job as a software developer at CERN in Geneva, Switzerland. He'd left some of his notes at home in England. Wouldn't it be great, he thought, if there was a piece of software that kept track of all the details in all his documents "that brains are supposed to be so good at remembering but that sometimes (his) wouldn't"? There wasn't such a program—so he wrote it himself and called it "Enquire." Of course, it only worked on his own personal files.

But Berners-Lee could see that Enquire had possibilities beyond that. He envisioned a system of open documents, all written in a common language and all linked together...which turned out to be those underlined words or phrases that we click on to take us to a different page or site.

He wrote a coding system called HTML (hypertext markup language); came up with an addressing system that gave each file on his "web" a unique address, which he called a URL (uniform resource locator); and wrote a program that allowed documents with a URL to be accessed by computers across the Internet: HTTP (hypertext transfer protocol). Finally, Berners-Lee took the step that would bring the whole world to what he would name the World Wide Web—the browser. By 1996, the number of Internet users had hit 40 million. At one point the rate of users was doubling every 53 days.

Berners-Lee himself hasn't profited at all from his creations. He manages the nonprofit W3 Consortium (which oversees development of web technology standards) from a plain, tiny office at MIT. The protocols he created are a household name—but Berners-Lee is content to keep working behind the scenes.

PROGERIA

This is an especially depressing disease that causes small children to physically age far more rapidly than normal. A young child afflicted with progeria can look 80 or 90 years old, and can already show symptoms regularly associated with old age, including wrinkled skin, hair loss, arthritis, loss of vision, and debilitated organ function. Progeria is caused by a mutation to a specific section of a specific gene, the full function of which scientists have yet to determine, although part of its function is obviously related to aging. (For that reason, the gene has been the subject of intense study by geneticists all over the world since its discovery in 2003.) Progeria is very rare, occurring in only about one in 8 million births. There is no known cure, and victims seldom live past their early teens.

PROJECT BLUE BEAM

Ever heard of Project Blue Beam? It's a conspiracy theory that, strangely enough, may have been inspired by a Star Trek *movie that was never made.*

In 1994 a biography was released called *Gene Roddenberry: The Myth and the Man Behind* Star Trek. The book told of a botched 1975 deal to put the canceled TV show on the big screen with a story about a flying saucer in Earth's orbit that sends down false gods in order to fool humans into servitude. No one had heard about that plot before. Not long after the biography was released, a Canadian journalist named Serge Monast started writing about a supposed secret plot called "Project Blue Beam" in which a shadow government masquerading as NASA fools humans into believing in aliens and gods. Monast said that there would be four steps to this dastardly plan:

1] The conspirators would set off a series of man-made earthquakes that would reveal (planted) archaeological discoveries proving all the major religions are false.

2] They would use top-secret holographic technology to make it appear as if gods and aliens were flying in the skies and in space.

3] "Telepathic Electronic Two-Way Communication" would be projected into every human's brain in which the "antichrist" will inform humanity about a glorious "New Age" to come (using some other type of advanced technology).

4] Finally, they would stage a simulated space battle/rapture, during which thousands of people would be whooshed up into the sky (using still more top-secret technology).

After that, the theory goes, the human race would fall in line with the New World Order. Even though Monast died from a heart attack in 1996 (or, as some maintain, he was murdered by the Canadian government), there are still many believers who claim that the ongoing Project Blue Beam can explain everything from an uptick in alien abduction reports to the 9/11 attacks to the holographic resurrection of Tupac Shakur at a 2012 concert.

Did Monast merely repeat a fictional *Star Trek* plotline, or did he unveil a nefarious conspiracy? You be the judge.

FIVE FREAKY FACTS ABOUT...
MICROWAVES

- No one knows why, but foods microwaved in a round container cook better than in a square one.
- In 1968 the Walter Reed Hospital tested microwave ovens to see if the waves (called microwaves) leaked out. They did—and the U.S. set the first federal standards for microwave ovens.
- Irregularly shaped foods, such as a leg of chicken that is thick at one end and thin at the other end, cook unevenly. That's because the microwaves penetrate completely through smaller pieces of food, but not through larger pieces.
- Aluminum foil reflects microwave energy the same way mirrors reflect light energy. That's why you can't use foil in a microwave.
- According to legend, shortly after Raytheon produced its first microwave oven in 1947, Charles Adams, the chairman of Raytheon, had one installed in his kitchen. But as Adams's cook quickly discovered, meat didn't brown in the oven, French fries stayed limp and damp, and cakes didn't rise. The cook, condemning the oven as "black magic," quit.

SEEING CLEARLY

Hans Lippershey was a 17th-century maker of eyeglasses. One morning, having just completed a pair of lenses, he stood in his shop doorway and inspected his work for imperfections. As a final test, he held both lenses up to the light and checked for minute flaws. What he saw next caused Lippershey to stagger back in amazement. Shaking his head in disbelief, he put the lenses up to his eyes once more. Again it happened—the church tower in the distance leapt out at him!

Hans had stumbled upon a way to make distant objects appear as if they were right in front of you. Lippershey had looked through two lenses at the same time—one concave (curved inward), the other convex (curved outward). Seeing a quick buck, he mounted the two lenses on a board and charged his customers to take a closer look at the distant church tower. After some experimentation, he mounted the lenses inside a hollow tube, dubbing the nifty device his *kijkglas* ("look glass").

We know it as the telescope.

THE MEANING OF LI-FI

*If you've ever tried to access a wireless router, you know
most Wi-Fi networks have pretty boring names...but
some people go for laughs when naming theirs.*

The LAN before time

Area 51

Your Bathroom Shower
Needs New Tiles

Help Me Pay For It

The Meaning of LiFi

Secret CIA Intelligence
Underground Military
Base

Why Phi?

Holy *$#% We're Online

The Dingo Ate My Wi-Fi

It Hurts When IP

Mom Click Here For
Internet

Alien Abduction Network

Global Thermonuclear War

SHUT YOUR DOG UP OR
I WILL CALL THE COPS

IP Freely

Caitlin stop using our
Internet!

HeyUGetOffMyLAN

.– .. ..-. .. (Morse code
for WiFi)

I'm Under Your Bed

Nuclear Launch Detected

Secret Federal Witness
Protection Safehouse

I'm cheating on my WiFi

John Wilkes Bluetooth

c:\virus.exe

Router—IHardlyKnowHer

Bill Wi the Science Fi

IfYouGuessMyPassword
IHaveToRenameMyDog

LAN of Milk and Honey

FECAL MATTERS

Since 1991, the Ig Nobel Awards honor the world's
most ridiculous research and scientific undertakings.
Here's one of our favorites.

The 2014 Ig Nobel award for biology went to a group of
scientists from the Czech Republic and Germany who
studied if dogs can sense the Earth's magnetic field. The
team examined dozens of dogs while they went to the
bathroom, theorizing that the average pooch aligns its body
along a north-south axis while they do so...or at least they do
when a solar storm isn't messing with the magnetic field.
After studying over 70 dogs and 1,893 "doggy dumps," they
published their findings in *Frontiers in Zoology*. (This same
group once conducted a similar study that proved that cows
also align themselves on an axis during times of relief,
but they studied that by examining Google Earth satellite
images.) Why do dogs and cows prefer to poop in this
fashion? The scientists have yet to figure that out.

IT'S ELEMENTARY

Remember the periodic table? Let's hope so...

1. Which extremely flammable element (atomic #1) was used to fill the *Hindenburg* instead of the element it was designed to use—helium?

2. The famed Anaconda mine in Butte, Montana, produced 94,900 tons of which element (#29) between 1881 and 1947?

3. Which element (#13), the most abundant metal in the earth's crust, is extensively used in compact discs, kitchen utensils, and house exteriors?

4. This metallic element (#22), strong as steel but 45 percent lighter, is primarily used in aircrafts and missiles, yet also finds its way into golf clubs and tennis rackets.

5. Which element (#36) is more widely known as the birthplace of Superman than as the rare gas used in specialized high-speed photographic flashlamps?

6. Can you name this common element (#6) that has an isotope that has been used for years to date archaeological finds?

7 What is the rare and short-lived element (#102) that is named after the inventor of dynamite?

8. Which element (#27) has been used for centuries to impart a permanent rich blue color into glass, porcelain, pottery, and enamels?

9. What is the second most abundant element (#14) on earth (exceeded only by oxygen) and is the principal ingredient of glass?

ANSWERS: 1. Hydrogen 2. Copper 3. Aluminum 4. Titanium 5. Krypton 6. Carbon 7. Nobelium 8. Cobalt 9. Silicon

THE MATRIX HAS YOU

"The Matrix is everywhere," Morpheus explains to Neo.
"It is all around us. You can see it when you look out your
window or when you turn on your television." What is the
Matrix? It's a computer simulation that we are all living in,
according to the cult movie, anyway. But what if we actually
are living in simulation? It's more likely than you think.

Leading the "Simulation Theory" is Nick Bostrom,
an influential Swedish philosopher who has advised tech
moguls and world leaders. In 2003 he published a paper
called "Are You Living in a Computer Simulation?" which
put forth this chilling scenario: In the future (if we survive
the present) we will reach a "posthuman" stage wherein
our descendants will use supercomputing power to create
detailed simulations of past civilizations. According to
Bostrom, "We are likely among the simulated minds rather
than among the original biological ones." If he's right, then
everything you are and everything you know is nothing
more than computer code. As Neo would say, "Whoa."

There's even some evidence to back this up. While researching String Theory, theoretical physicist S. James Gates discovered something quite remarkable: buried deep in the equations that explain how the universe works are "error protecting codes," series of ones and zeros that couldn't occur randomly. In fact, they're nearly identical to programs that browsers use to correct errors when transmitting data. "How could we discover whether we live inside a Matrix?" Gates asked at a science symposium in 2016. "One answer might be, 'Try to detect the presence of codes in the laws that describe physics.'" And now he has.

So are we living in a simulated universe—or one of millions of simulated universes, as Bostrom theorizes? If so, that might explain a lot, like reincarnation (your past life was a past simulation). But could we ever know for sure? As another philosopher, New York University's David Chalmers, noted, if we are part of a computer simulation, then "any evidence we could ever get would be simulated."

DUMB

PREDICTIONS

"Heavier-than-air
flying machines are
impossible."

—Lord Kelvin, president,
Royal Society, 1895

Odd Books

If Strange Science *isn't quite odd enough for you, here are some even weirder books to look for.*

THE TOOTHBRUSH: ITS USE AND ABUSE, Isador Hirschfield (1939)

THE ROMANCE OF LEPROSY, E. Mackerchar (1949)

SEX AFTER DEATH, B. J. Ferrell and D. E. Frey (1983)

AMERICAN BOTTOM ARCHAEOLOGY, Charles John Bareis and James Warren Porter (1983)

THE RESISTANCE OF PILES TO PENETRATION, Russell V. Allin (1935)

CONSTIPATION AND OUR CIVILIZATION, J. C. Thomson (1943)

MAKING IT IN LEATHER, M. Vincent Hayes (1972)

THE FOUL AND THE FRAGRANT: ODOR AND THE FRENCH SOCIAL IMAGINATION, Alain Corbin (1986)

WHY PEOPLE MOVE, Jorge Balán (1981)

HANDBOOK FOR THE LIMBLESS, Geoffrey Howson (1922)

ETERNAL WIND, Sergei Zhemaitis (1975)

THE ROMANCE OF RAYON, Arnold Henry Hard (1933)

WHAT TO SAY WHEN YOU TALK TO YOURSELF, Shad Helmstetter (1982)

HISTORIC BUBBLES, Frederic Leake (1896)

HOW TO FILL MENTAL CAVITIES, Bill Maltz (1978)

A DO-IT-YOURSELF SUBMACHINE GUN, Gérard Métral (1995)

NUCLEAR WAR: WHAT'S IN IT FOR YOU? Ground Zero War Foundation (1982)

H. G. Wells, Futurist

In the final article in our series about the futurists, we examine what the future has in store...according to one sci-fi author.

British novelist H. G. Wells witnessed significant change in his lifetime. When he was born in 1866, cities were lit by torches and oil lamps, and there were no horseless carriages or air travel. By the turn of the century, cities were being lit by gas lamps, and automobiles were steadily replacing the horse. In 1901 Wells published his groundbreaking treatise on the future, *Anticipations*. In it, he foresaw the end of the steam age and the rise of oil. He accurately predicted that the entire region from Boston to Washington, D.C., would become one long system of suburbs, cities, highways, and traffic jams. He even predicted speed limits.

Yet for all his foresight, Wells got a lot wrong: He said that airplanes were just a passing fad and that moving sidewalks would be commonplace in cities. He also predicted that the world's governments would merge into one "New Republic" ruled by scientists who would eliminate all but the white race and "establish a world state with a common language and a common rule." That future hasn't arrived.

The End

Talk about alarmist headlines. In March 2014, the *Guardian* ran this whopper: "Nasa-funded study: industrial civilization headed for 'irreversible collapse'?" Similar headlines followed. The study in question was led by University of Maryland applied mathematician Safa Motesharri, and through it was financed in part by a grant from NASA, no one from NASA participated in the study. And they made that clear, stating, "NASA does not endorse the paper or its conclusions."

Motesharri's goal was to create a "universal model of social collapse" to explain why so many thriving civilizations—the Romans, the Mayans, the Mesopotamians—suddenly went kaput. He came up with four equations (full of strange symbols and Greek letters) that defined humans as "Predators"—divided into "Elites" and "Masses"—and nature as simply "Prey." Because nature can't be quantified, damage is measured in "Eco-dollars." The study's conclusion: "Given economic stratification, collapse is very difficult to avoid and requires major policy changes, including major reductions in inequality and population growth rates."

The good news: a whole bevvy of statisticians and scientists took Motesharri to task for his methodology (he provided no empirical evidence) and his overgeneralizing of complex systems. The bad news: Motesharri still might be right.

Index of Stories